Battery Swapping for Sustainable Transportation

Battery Swapping for Sustainable Transportation

Xiao Lin
Xue Wang
Bin Wu
Yingying Wang
Shuoyuan Zhang

WILEY-VCH

Authors

Xiao Lin
Botree Recycling Technologies, Room 202, Building 10, No. 99 Jinjihu, Avenue, Suzhou, Jiangsu, 215128, China

Xue Wang
Botree Recycling Technologies, Room 202, Building 10, No. 99 Jinjihu, Avenue, Suzhou, Jiangsu, 215128, China

Bin Wu
Botree Recycling Technologies, Room 202, Building 10, No. 99 Jinjihu, Avenue, Suzhou, Jiangsu, 215128, China

Yingying Wang
Botree Recycling Technologies, Room 202, Building 10, No. 99 Jinjihu, Avenue, Suzhou, Jiangsu, 215128, China

Shuoyuan Zhang
Botree Recycling Technologies, Room 202, Building 10, No. 99 Jinjihu, Avenue, Suzhou, Jiangsu, 215128, China

Cover Design: Wiley
Cover Image: © Chesky/Shutterstock

Typesetting: Lumina Datamatics, Madurai, India

Library of Congress Card No.: applied for

British Library Cataloguing-in-Publication Data
A catalogue record for this book is available from the British Library.

Bibliographic information published by the Deutsche Nationalbibliothek
The Deutsche Nationalbibliothek lists this publication in the Deutsche Nationalbibliografie; detailed bibliographic data are available on the Internet at <http://dnb.d-nb.de>.

Print ISBN 9783527354863
ePDF ISBN 9783527850983
ePub ISBN 9783527850976
oBook ISBN 9783527850990

Printed and bound by CPI Group (UK) Ltd, Croydon, CR0 4YY
C9783527354863_230626

Contents

About the Authors

Xiao Lin, PhD, is Founder and CEO of Botree Recycling Technologies. He received his doctorate from Chinese Academy of Sciences and focuses on battery material regeneration technologies and solutions. He is a member of SAC/TC294, Convenor of ISO/TC 333/AHG2, Co-chair of GBA CCMAG Steering Committee, Member of GBA Battery Passport Steering Committee, Operation Agent of IEA HEV TCP Task 48 Battery Swapping, participant of IEA HEV Task 40 CRM4EV.

Xue Wang, PhD, is COO and Principal Analyst of Botree Recycling Technologies. She obtained her doctorate in Material Engineering from KU Leuven. Her research is focused on new metallurgical process technology and waste resource treatment technology, and she is co-manager of IEA HEV-TCP Task54 Recycled Materials for Electric Vehicles.

Bin Wu is director of Sustainability Consulting at Botree Recycling Technologies, holding a master's degree in material science from KU Leuven. As an expert in product life cycle assessment (LCA), he has accumulated years of experience in product carbon footprint research at leading international manufacturing enterprises, with solid expertise in carbon footprint modeling and analysis. Currently, he is deeply engaged in research and analysis of carbon footprint, recycled content, regulation compliance across the lithium battery and new energy industry chains, and has participated in the compilation and review of numerous domestic and international carbon footprint standards, including participating in the expert review of the European Union Joint Research Centre (JRC)-led battery regulation carbon footprint calculation rules (CFB-EV for electric vehicle batteries, CFB-INB for industrial batteries), contributing to the formulation of automobile life cycle assessment regulations under the United Nations World Forum for Harmonization of Vehicle Regulations (UN/WP.29), the global battery alliance (GBA) battery greenhouse gas (GHG) calculation rulebook, and the international lithium association (ILiA) lithium product carbon footprint guidelines. He has undertaken over 50 sustainability-related consulting projects.

Yingying Wang is a master's graduate of Southwest University and a Senior Engineer of the Chinese Academy of Sciences. She currently serves as senior manager at Botree Recycling Technologies, China, with core expertise in carbon trading mechanisms, lithium battery industry chain sustainability, and ESG compliance. She is a member of the Jiangsu Provincial Metrology Technical Committee for Carbon Peak & Carbon Neutrality, and an expert at the Suzhou Industrial Park Quality Infrastructure "One-Stop" Service Center (New Energy & Green Industry). With 10+ years of experience in climate change, ESG, and environmental consulting, she has served industry leaders including BMW, Huawei, CATL, and TQC.

Shuoyuan Zhang earned his master's degree from the University of Glasgow in the United Kingdom, majoring in International Economics and Finance. His research interests focus on economics and industrial finance, mainly concentrating on the new energy sector, including photovoltaics, charging, and battery swapping industries.

Preface

We stand at an inflection point in the reshaping of transportation energy systems. If the first stage of electrification can be summarized as "replacing the engine with a battery," the next stage is defined by "turning the battery into a node of the energy internet." In this profound transformation, battery swapping technology, with its unique systemic value, emerges as a critical nexus connecting vehicles, batteries, the power grid, and renewable energy sources, offering a vital and alternative pathway toward a sustainable transportation future.

As an efficient mode of energy replenishment, the significance of battery swapping extends far beyond "faster charging." Through the physical circulation of batteries, it decouples the vehicle from its energy source, catalyzing business model innovations like "vehicle-battery separation," which lowers the upfront cost barrier for users. It enables centralized, controlled slow charging and refined management of battery packs, significantly extending battery lifespan while enhancing safety and resource utilization efficiency. More importantly, it unlocks the potential for massive fleets of vehicle batteries to function as scalable, dispatchable distributed energy storage units, actively participating in grid peak shaving and valley filling, thereby facilitating the integration of intermittent renewable energy sources like wind and solar power. Thus, battery swapping evolves from a mere technical option into a systemic endeavor encompassing technology, business, energy, and policy.

This book provides a comprehensive, multi-dimensional analysis of this evolving system. Chapter 1 traces the developmental trajectory of the global and Chinese battery swap industry, setting the context within the broader landscape of transportation electrification. Chapter 2 assesses the role of battery swapping in the global energy transition, evaluating its potential for lifecycle emission reduction and its contribution to resource efficiency through standardized design and extended battery use.

Chapter 3 delves into the core of technological innovation and standardization, examining global research trends, patent dynamics, and the critical challenges in establishing interoperable technical standards. Chapter 4 forms the empirical heart of the volume, offering a detailed analysis of business models and case studies across diverse applications—from electric two/three-wheelers and passenger vehicles to commercial trucks, and extending to emerging fields like maritime and electric aviation.

Chapter 5 introduces and analyzes the "Battery Bank" model, a key financial and operational innovation underpinning the swap ecosystem. Chapter 6 systematically maps the entire industry chain, detailing the strategic roles of upstream suppliers, midstream operators and manufacturers, and downstream service providers. Finally, Chapter 7 synthesizes the prevailing challenges related to infrastructure, cost, safety, and policy, while outlining the future opportunities and developmental prospects for the industry.

For policymakers, this book aims to serve as a "decision-making sandbox," presenting a dynamic comparison of battery swapping versus charging, hydrogen energy, and synthetic fuels in terms of cost curves and carbon-reduction efficiency; for entrepreneurs, we seek to break down the key resource capabilities across different sectors—the density economy of two-wheeler battery swapping, the route economy of heavy-duty truck battery swapping, and the subscription-based flexibility of passenger vehicle battery swapping; for investors, "Battery Bank DCF Model" and "Battery Swap Station REITs Case Study" may offer new valuation perspectives.

Yet battery swapping is not without contention: capital-intensive infrastructure, fragmented standards, second-life battery safety, grid impact, and the profitability of business models all challenge the sector to deliver superior solutions. The final chapter proposes a four-dimensional "technology–standards–policy–market" synergy framework and a phased roadmap for 2025–2035, offering an actionable checklist for corporate strategists, investors, and policymakers.

Acknowledgments to the International Energy Agency Electric Vehicle Technology Collaboration Programme (IEA EV TCP) and all the experts and participants of IEA EV TCP Task 48 Battery Swapping. Acknowledgments to Wenjun Tang, Guoqing Chen, Yong Zeng, Yuan Yuan, Fan He from Wuhan Changjiang Ship Design Institute Co., Ltd., Liguo Li, Yunchuan Zeng, Dan Liu from Sichuan Zhili Intelligent Energy Technology Co., Ltd., Changwei Cai from Suzhou Harmontronics Automation Technology Co., Ltd, Dongdong Cao, Baoqiang Zhang, Zhanhui Yao, Yunzhe Jiang from China Automotive Technology and Research Center Co., Ltd., Chao Yang, Ning Zhang, Yingxin Wang from Wuhan NIO Energy Co., Ltd, Kai Zhang from Contemporary Amperex Technology Co., Ltd., Syb ten Cate Hoedemaker from Maritime Battery Forum, Sherry Chen from Nickel Institute, Bingzhang Chen, Lei Cao, Jiawei Wen from China University of Petroleum(Beijing), and Shengrui Qi, Ilai Bendavid and Tina Wu from Botree Recycling Technologies.

If the first stage of electrification was about "replacing the engine with a battery," the next stage is about "turning the battery into a node of the energy internet." At the heart of this shift, battery swapping is the key to this future. May this book serve as yours—to understand, engage, and shape this transformation.

1

Introduction

1.1 Current Landscape of Transportation Electrification

Amid significant shifts in global energy systems and increasingly stringent environmental regulations, the transition to electric transportation is gaining momentum at an unprecedented pace. From electric two-wheelers (E2W) to electric three-wheelers (E3W), these short-distance travel tools have achieved notable success, particularly in Asia, where they have become a vital force in urban commuting and last-mile deliveries, reducing pollution and conserving energy.

Worldwide, the electrification of passenger vehicles has seen remarkable progress. Industry pioneers, exemplified by Tesla along with a growing number of automakers, are introducing high-performance electric vehicles, consistently improving range, shortening charging durations, and extending battery longevity. Plug-in hybrid electric vehicles (PHEVs) and hydrogen fuel cell vehicles (FCVs) are also advancing, providing diverse options for consumers.

Commercial vehicle electrification is also progressing steadily. Pilot programs in buses, logistics vehicles, and sanitation vehicles have been initiated, with many countries setting clear goals to increase the proportion of new energy vehicles in public fleets and to phase out fossil-fueled commercial vehicles. In China, by 2025, the electrification of public transportation will be further accelerated. Not only will urban bus systems be almost entirely electrified, but rail and waterway freight will also introduce more electric and hydrogen-powered equipment, aiming to significantly reduce carbon emissions in the transportation sector.

Electric ships and aircraft are also showing strong trends toward electrification. For instance, battery giants like CATL are applying their advanced battery technology to ship and aircraft manufacturing, marking a significant expansion of battery applications beyond land-based transportation to include maritime and aerospace sectors, achieving comprehensive electrification from land to sea and sky. The development and deployment of electric ships help reduce pollution in ports and inland waterways, while electric aircraft promise a new era of green aviation.

As the electric vehicle (EV) market grows rapidly, efficient and convenient energy replenishment has become a critical bottleneck. With improvements in battery energy density and cost reductions, the increased driving range of EVs has heightened consumer expectations for faster recharge times. While slow charging meets basic needs, longer wait times in long-distance driving and in frequent-use scenarios pose challenges to user experience.

As an alternative energy replenishment method, battery swapping enables quick battery replacement, drastically cutting down waiting times and addressing concerns over driving range. For example, NIO's BaaS (Battery as a Service) enables users to swap batteries at standardized swap stations within minutes, effectively restoring their vehicles to full charge.[1]

In the domain of commercial vehicles, including electric buses and trucks, the demand for substantial cargo capacity and continuous operation renders battery swapping particularly advantageous, especially considering the size and weight of their batteries.[2] Automated battery swapping utilizing specialized equipment not only enhances safety but also ensures fleet efficiency while minimizing downtime costs associated with charging. In the micro-electric vehicle sector, such as electric motorcycles and tricycles, self-service battery swapping has gained mainstream acceptance due to the simplicity of vehicle design and smaller battery dimensions. Service providers have established dense networks of swap cabinets that enable users to effortlessly replace pre-charged battery modules, thereby enhancing convenience.

Globally, battery swapping is increasingly receiving policy support and market acceptance. Beyond easing the burden on conventional charging networks, it enhances energy management and operational efficiency, fostering the sustainable growth of the EV sector.[3] However, large-scale implementation faces challenges related to battery standardization, network configuration, and investment return periods.

Through systematic research, initiatives like Battery Swapping for Sustainable Transportation aim to equip policymakers with scientific insights and detailed decision-making frameworks that support the formulation of policies conducive to advancing electric transportation. Additionally, it provides strategic guidance for businesses navigating the competitive landscape of electrification while contributing to an overall enhancement of the EV value chain—ensuring environmental protection alongside sustainable socioeconomic development.

---

1 [The Role of Battery Swapping in Accelerating Electric Vehicle Adoption], John Doe, Jane Smith, 2022.

2 [Technological Innovations in Battery Swapping for Commercial Electric Vehicles], David Lee, Sarah Wang, 2023.

3 [Policy Support for Electric Vehicle Infrastructure: A Global Perspective], Michael Johnson, Emily Brown, 2021.

1.2 Overview of the Swapping Industry Development

In 2007, Shai Agassi pioneered battery swapping technology for EVs in Israel by founding Better Place, a company that successfully raised $850 million in investments. One year after its establishment, it constructed its first battery swap station in Israel and entered into a cooperation agreement with Renault-Nissan to provide 100,000 Fluence Z.E. EVs. However, the company faced significant challenges due to the high investment costs associated with swap stations, the depreciation of battery assets resulting from advancements in battery technology and cost reductions, as well as low market acceptance (with only 1% of actual orders). Consequently, Better Place struggled to maintain its operations and ultimately declared bankruptcy in 2013. In the same year, Tesla developed a quick-swap technology capable of completing swaps in just 90 seconds; however, it encountered similar issues, such as high costs, low consumer acceptance rates, and poor compatibility. As a result, Tesla abandoned the battery swap model merely two years later.

1.2.1 China's Battery Swapping Evolution

The evolution of battery swapping in China can be categorized into three distinct stages:

2006–2011: Technology Preparation and Business Model Exploration Phase

In 2006, the State Grid initiated research and development for EV charging and battery swapping infrastructure.

In 2010, a pilot project was conducted in Hangzhou using Zotye Langyue and Haima Prima electric taxis equipped with high-capacity battery boxes, which tested the "vehicle-battery separation and mileage-based charging" business model.[4]

In 2011, the State Grid established its operational model for smart charging and battery swapping, prioritizing "battery swapping as the primary method, plug-in charging as a secondary option, along with centralized charging and uniform distribution."

2012–2018: Focus on Charging Mode Development Phase

Following the release of the "Energy-Saving and New Energy Vehicle Industry Development Plan (2012–2020)" in 2012, China identified charging infrastructure as a primary development focus.

During this period, the State Grid's battery swap projects were suspended due to challenges in integrating the entire industry chain, high investment costs, and strategic realignments.

---

[4] Liang, Y., Zhang, X. J. E., Battery swap pricing and charging strategy for electric taxis in China. *Energy* 2018, 147, 561–577.

Despite the focus on charging, automakers continued to develop battery swapping technologies. BAIC New Energy, together with AutoNavi and Shanghai Diba, expanded battery swapping operations from Beijing to the rest of the country. In 2017, BAIC launched the "Titan" plan.

In 2018, NIO introduced the ES8 featuring a high-speed battery swap mode and implemented a "vehicle-battery separation" model.

2019–Present: Rapid Development Phase of Battery Swapping

The year 2020 marked a milestone for battery swapping as it was recognized as a critical component of new infrastructure and prominently featured in China's 'Government Work Report.'

Subsequently, the government launched a series of policies encouraging the application of battery swapping models and officially started the pilot program for battery swapping in the new energy vehicle industry in 2021.

1.2.2 Key Challenges in the Initial Global Trajectory

Reflecting on the global trajectory of battery swapping, several key factors contributed to its initial challenges:

- The limited market scale and low ownership rates of vehicles during the early stages of the new energy vehicle industry made it difficult to support large-scale construction and operation of battery swap stations.
- High initial investments and service fees resulted in low consumer acceptance and utilization rates at swap stations, thereby prolonging asset recovery cycles and impeding sustainable economic returns.
- The absence of a comprehensive technical standard system encompassing both battery swapping technology and related infrastructure led to inconsistencies and a lack of unified norms.
- Compatibility issues between vehicle interfaces and batteries posed safety risks while diminishing user convenience.
- Insufficient value distribution mechanisms between automakers and operators of battery swap stations hindered collaboration in resource sharing, cost-bearing, and benefit distribution.

In summary, the initial challenges encountered by battery swapping on a global scale primarily stemmed from economic factors, technical standardization issues, and the integration of industry value chains. As the new energy vehicle market rapidly expands, advancements in technology and policy-driven efforts toward standardization are effectively addressing these challenges. This progress is paving the way for a new phase of development in battery swapping.

1.2.3 Global Resurgence of Battery Swapping

The global resurgence of battery swapping in recent years has been remarkable. It serves as a pivotal driver of innovation within the EV sector and represents a strategic pathway toward low-carbon transportation through energy system transformation.

Particularly in China, the thriving new energy vehicle industry, supported by national policies and driven by market demand, has increasingly highlighted the value and advantages of battery swapping.

Chinese industry leaders like NIO have successfully implemented battery swapping services for passenger vehicles, establishing an extensive network of swap stations that enable users to replace batteries in just a few minutes. This development has significantly alleviated concerns regarding EV range and lengthy charging waiting times, thereby enhancing user experience and acceptance. Furthermore, the business model—which encompasses battery leasing and subscription services—lowers the initial purchase cost for consumers while stimulating market activity.

In the commercial vehicle sector, particularly within logistics and public transit, battery swapping has fundamentally transformed operational models. Due to the high frequency and intensity of usage, traditional slow charging methods are inadequate for meeting continuous operational demands. Battery swapping, characterized by its efficiency and rapidity, effectively addresses long-standing issues such as "difficult charging" and "extended waiting times," significantly enhancing vehicle uptime and overall operational efficiency. This advancement accelerates the electrification process for commercial vehicles.

With ongoing technological innovation and cross-sector integration, battery swapping technology is expanding into broader transportation domains. Electric ships and electric aircraft are beginning to implement battery swapping or similar concepts to address energy replenishment challenges, particularly in short-range, high-frequency ferry and drone operations. The development of modular, quickly replaceable battery packs has become a focal point of research, highlighting the potential for battery swapping to disrupt additional transportation scenarios while demonstrating adaptability across various applications.

The rapid global advancement of battery swapping not only optimizes travel experiences for current EV users but also stimulates diverse business models and technological innovations across multiple market segments. This progress contributes significantly to achieving global green transportation goals and reducing emissions. As technology continues to mature alongside market dynamics, battery swapping is poised to play a pivotal role in mainstreaming new energy vehicles by leveraging its unique advantages across more transportation sectors, ultimately generating greater social and economic benefits.

1.3 Scope of Research and Main Research Contents

(Two- and Three-wheelers, Passenger Cars, Commercial Vehicles, Electric Ships, Electric Aviation)

The research scope of this book spans electric two- and three-wheelers, passenger cars, commercial vehicles, and emerging fields like electric ships and aviation, with a primary focus on battery swapping.[5] The main contents include:

[5] [Battery Swapping for Electric Vehicles: A Comprehensive Review of Technologies, Policies, and Market Potential], Zhang Wei, Li Ming, 2021.

An extensive analysis of market demand for battery swapping across different EV segments is conducted, highlighting the benefits of this model compared to conventional charging and pinpointing key use cases.

The book systematically examines the policy landscape, market dynamics, and technological progress in the battery swapping industry globally and in China, tracing the evolution of the model and showcasing successful implementations.[6]

Critical areas such as infrastructure development, business model innovation, battery asset management, and standardization are explored in depth, addressing key challenges and advancements.[7]

The adaptability and potential of battery swapping in electric ships and aviation are analyzed, taking into account the unique requirements of these sectors, with projections for future developments.

Based on real data, we scientifically predict the market size of battery swapping in various sub-sectors over the coming years and analyze the potential opportunities and challenges.

By thoroughly examining the research topics, this book offers a detailed and multi-dimensional perspective on the battery swapping sector, fostering ongoing innovation and promoting high-quality industry growth.

---

6 [Policy Drivers and Market Dynamics of Battery Swapping in China], Chen Xia, Wang Lei, 2022.

7 [Battery Swapping Infrastructure: Challenges and Opportunities for Standardization], John Smith, Emily Brown, 2023.

2

The Impact of Battery Swapping on Global Energy Transition

2.1 Sustainable Development Goals and Carbon Neutrality Pathways

The global energy sector accounts for ~75% of GHG emissions, driving urgent transitions to net-zero systems by mid-century. The IEA highlights that limiting warming to 1.5 °C requires intensified international collaboration. At COP28 (2023), nearly 200 nations pledged to triple renewable energy capacity by 2030 and accelerate fossil fuel phase-out. A streamlined overview of key economies' carbon neutrality strategies has been summarized in Table 2.1:

Major Economies' Carbon Neutrality Targets and Initiatives

Table 2.1 Global climate targets and key policies by region.

Region	2030 target	2060 target	Key policies and actions
EU[a]	55% emissions reduction (vs. 1990)	Climate neutrality	• European Green Deal: 42.5–45% renewables by 2030 • REPowerEU Plan: Accelerate clean energy investment • Green Deal Industrial Plan: 40% domestic net-zero tech production by 2030[b]
US	50% EV sales share	Carbon neutrality	• Inflation Reduction Act: $369B for EVs/batteries, North American raw material mandate • Bipartisan Infrastructure Law: $1.2T for clean energy infrastructure

(Continued)

Table 2.1 (Continued)

Region	2030 target	2060 target	Key policies and actions
Japan	100 Mt emissions reduction (via JCM mechanism)	Carbon neutrality	• Green Growth Strategy: ¥2T fund for hydrogen/CCUS • 420 municipalities committed to 2050 neutrality
China	Carbon peaking	Carbon neutrality (2060)	• Dual Carbon Strategy: 25% nonfossil energy by 2030 • National carbon trading market expansion • "PV-storage-charging-swapping" pilot projects[c]
Southeast Asia	Enhanced renewable energy integration	N/A	• Regional cooperation on clean energy • Reduced single-use plastics, improved public transport
India	45% GDP emissions intensity reduction (vs. 2005)	Net-zero (2070)	• National Solar Mission: 280 GW solar capacity by 2030 • EV infrastructure expansion • Energy efficiency schemes (e.g., PAT)

a. REPowerEU (europa.eu).
b. Green Deal Industrial Programme - European Commission (europa.eu).
c. China's Climate Change Policies and Actions 2023 Annual Report.

2.2 The Trend from Fossil Fuels to Electric Vehicles

The electrification of transport, particularly road transport, is widely recognized as the most significant and effective strategy for reducing emissions in the transportation sector. This perspective has been embraced by numerous countries, economies, and automobile manufacturers worldwide. It happened with a faster pace than people expected. The International Energy Agency (IEA) has predicted in its early report that if the future targets for electric vehicles (EVs) from the ten largest car manufacturers globally have been achieved, the annual global sales of EVs could surpass 20 million units by 2030. Nevertheless, this target is likely to be accomplished this year in 2025. Taking into account the ambitions set forth by other automotive companies, it is projected that the electrification rate of cars sold worldwide may exceed 50% by 2030.

The history of EVs dates back to the nineteenth century, predating the advent of internal combustion engine (ICE) cars. In 1834, American inventor Thomas Davenport created the first electric tricycle powered by dry batteries. This innovation paved the way for large-scale commercial applications in the late nineteenth and early twentieth centuries. However, as ICE technology rapidly advanced during the mid-twentieth century, EVs were largely supplanted by fuel-powered alternatives.

It was not until 2010 that EVs began to experience a resurgence, driven by heightened environmental awareness and technological advancements. The twenty-first century has heralded new development opportunities for this sector. Particularly since 2019, governments worldwide have set ambitious carbon neutrality goals and implemented policies to support the growth of EVs. Concurrently, significant breakthroughs in battery technology and a marked reduction in costs have led to substantial improvements in range, charging times, and overall economic viability of EVs—resulting in explosive growth within this market.

According to recent data released by IEA, EV sales reached nearly 17 million globally in 2024, as can be seen in Figure 2.1, reflecting a penetration rate of more than 20% and bringing total ownership to around 58 million units.

Despite the rapid overall growth of EVs in recent years, the current stage of development exhibits several notable features:

High Regional Concentration with significant increase in emerging markets: In 2024, China (65.3%), Europe (17.6%), and the United States (9.4%) are still the three major markets, accounting for 92.4% of global EV registrations, significantly higher than the proportion of ICE vehicles during the same period. Remarkably, outside of these three major markets, there was a record increase in sales of nearly 40% to reach 1.3 million.

Rapid Increase in Available Models, Primarily SUVs and Large Vehicles: The number of EV models offered by automakers is increasing at an annual compound growth rate of 15%, reaching 785 in 2024, and is projected to match ICE vehicle offerings by 2030. Among the available models, there is a clear trend toward SUVs and larger vehicle types, which can be clearly seen in Figure 2.2.

Price gap keeps narrowing: The price gap between EVs (EVs) and ICE vehicles is gradually narrowing, mainly thanks to the price drop of battery packs. The global average battery pack price fell more than 25% compared with 2023 levels, as can be read from Figure 2.3. This resulted in a global drop in electric car manufacturing costs that was reflected in the price of electric cars.

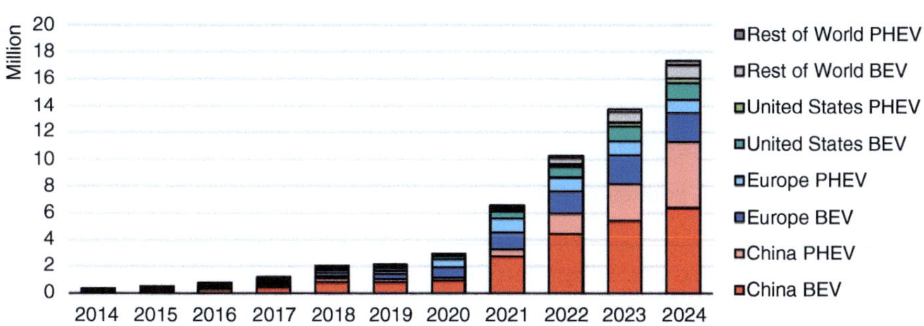

Figure 2.1 Global electric car sales, 2014–2024.

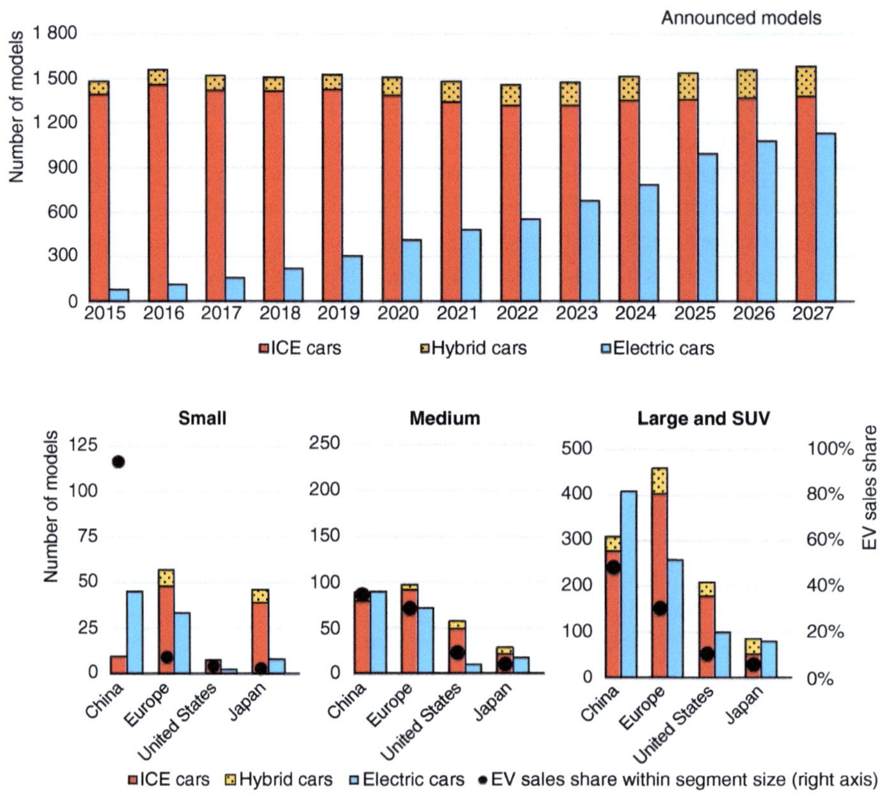

Figure 2.2 Global trends in the share of optional EV models and SUVs, and large vehicles.

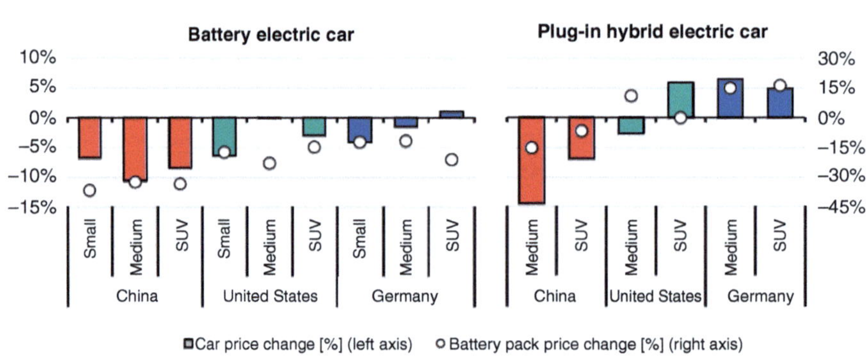

Figure 2.3 Electric car price and battery system price changes in selected countries.

Low penetration of electric bus outside China and Europe: Electric bus adoption remains relatively low in most regions outside of China and Europe (Figure 2.4). On a global scale, electric buses accounted for around 5% of total bus sales in 2024. However, in certain countries such as China, Belgium, Norway, and Switzerland, the penetration rate of electric buses has exceeded 50%.

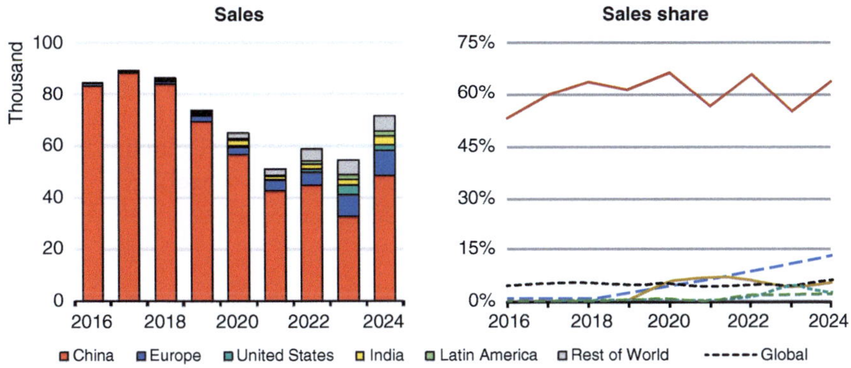

Figure 2.4 Electric bus sales and penetration rate trend.

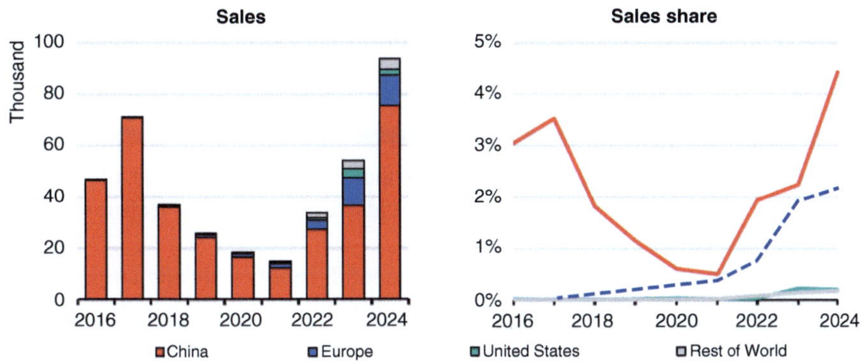

Figure 2.5 Electric truck sales and penetration rate trend.

Truck electrification just beginning: The electrification of trucks is in its nascent stages, with overall penetration remaining minimal. As can be seen from Figure 2.5, in 2024, global sales of medium- and heavy-duty trucks grew for the third consecutive year in 2024 to exceed 90,000 worldwide, with almost 80% growth-on-year. The overall level of truck electrification continues to be low, even in China, which holds the title as the world's largest market for electric trucks, the penetration rate has never exceeded 5%. The widespread adoption of electric trucks still faces significant challenges ahead.

2.3 Role of Battery Swapping in Reducing Greenhouse Gas Emissions

When assessing the emission reduction potential and trends associated with EVs (EVs), a lifecycle analysis approach is commonly utilized. For comprehensive information on lifecycle analysis methodologies, please refer to the ISO 14040/44 standards and relevant literature, which will not be addressed in this context.

The lifecycle carbon emissions of a vehicle comprise two critical components: the vehicle cycle (VC) and the fuel cycle (FC).

Vehicle cycle: This includes the production of automotive raw materials and parts, vehicle assembly, maintenance (such as tire and battery swapping), and end-of-life disposal.

Fuel cycle: Also known as the well-to-wheel (WTW) cycle, this encompasses the production and transportation of fuel (such as gasoline or electricity) from the source (well-to-tank) to its utilization in the vehicle (tank-to-wheel).

These components are illustrated in Figure 2.6.

2.3.1 The Potential Impact of Battery Swapping on Greenhouse Gas Emissions

EVs play a pivotal role in mitigating greenhouse gas emissions and possess significant potential to impact the environment and address climate change. Firstly, the zero-emission feature of EVs renders them an environmentally sustainable option for road transportation, as they operate on battery power rather than conventional fossil fuels, thereby eliminating exhaust emissions during operation. Secondly, when utilized in conjunction with renewable energy sources such as wind and solar power, EVs can further diminish their greenhouse gas emissions throughout their operational lifespan. The advancement of EVs catalyzes innovation and progress in related fields such as battery technology and charging infrastructure.

Battery swapping enhances EVs environmental benefits through operational efficiency (detailed in Table 2.2), extended battery life (detailed in Table 2.3), and standardized recycling (detailed in Table 2.4). Below is a structured overview of its key impacts:

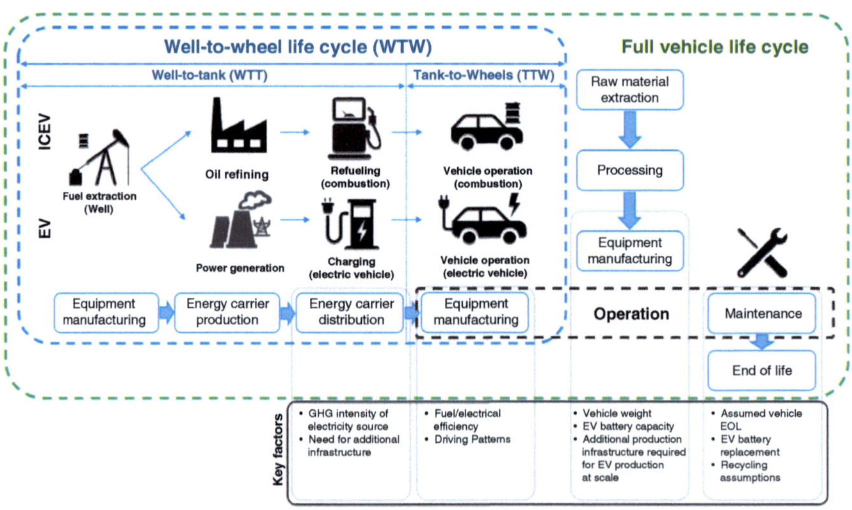

Figure 2.6 The boundaries of the whole life cycle system of the vehicle.

Direct Emission Reductions via Operational Efficiency

Table 2.2 Comparison of traditional charging and battery swapping models.

Aspect	Traditional charging	Battery swapping model	Impact data
Energy consumption	Higher reliance on grid energy (often fossil fuel-based)	Optimized energy use via centralized, renewable-powered swap stations	• Haikou taxi pilot: 0.01 gce/MWh lower energy consumption
Emission source	Indirect emissions from grid electricity	Reduced grid dependency; integrates solar/wind power	EVs with PV-powered swapping achieve 75% lower emissions vs. traditional ICE vehicles

Extended Battery Lifespan and Reduced Carbon Footprint

Table 2.3 Comparison of traditional charging and battery swapping models: mechanism and benefits.

Mechanism	Traditional charging	Battery swapping model	Quantifiable benefits
Charge–Discharge cycles	Deep cycles (DOD > 70%) cause rapid degradation	Shallow cycles (DOD ≤ 50%) minimize wear	Battery lifespan extended from 7.7 to 17 years • Reduces manufacturing-related emissions by 55%
Manufacturing impact	Frequent replacements increase resource use	Reduced replacement frequency lowers raw material demand	Standardized designs cut production costs by 20% and recycling costs by 30%

Environmental Benefits of Standardized Design

Table 2.4 Comparison of processes: non-standardized vs. standardized swapping batteries.

Process	Challenges with non-standardized batteries	Advantages of standardized swapping batteries
Production	Higher defect rates, inconsistent energy use	Uniform manufacturing reduces waste (e.g., CATL's "Chocolate Block" design)

(Continued)

Table 2.4 (Continued)

Process	Challenges with non-standardized batteries	Advantages of standardized swapping batteries
Recycling	Complex disassembly, low material recovery	Modular design enables >95% electrode material recovery (vs. 70% in traditional methods)
Grid integration	Fragmented energy storage capabilities	Swap stations act as distributed energy storage nodes (V2G/VS2G), reducing peak load by 15%

The battery swapping model has emerged as a pivotal technological pathway for EVs to achieve deep decarbonization, leveraging a full-chain mechanism that integrates operational emission reduction, battery lifespan optimization, and energy system synergy. This multi-dimensional approach not only addresses the immediate emission challenges during EV operation but also systematically tackles the embedded carbon in battery production and grid dependency, thereby creating a closed-loop sustainability ecosystem.

2.3.2 Battery Swapping Reduces GHG Emissions via Extended Battery Lifespan

By reducing deep charge-discharge cycles, the lifespan of the battery can be effectively extended, improving its cycle stability and reliability. As illustrated in the Figure 2.7, as battery lifespan increases, the impact of greenhouse gas emissions from battery manufacturing diminishes across various usage scenarios in different countries.

Battery swapping presents an innovative solution to the challenges faced in EV usage. While this model does not mitigate carbon emissions during the manufacturing phase of power batteries, it significantly enhances their utilization efficiency. According to data from relevant automobile manufacturers, under the battery swapping model, power batteries typically retain approximately 80% of their original capacity after traveling between 300,000 km and 450,000 km (China Automotive Low Carbon Action Plan 2020).

Moreover, due to the absence of time constraints at battery swap stations, these facilities generally employ slow charging methods for centralized charging. This approach ensures greater consistency among batteries and effectively mitigates potential damage to battery life that can result from fast charging practices. Research indicates that carbon emissions associated with power batteries can account for up to 40% of the total lifecycle emissions of EVs. Consequently, as battery lifespan increases alongside mileage traveled, carbon emissions per kilometer under the battery swapping model are expected to decrease progressively with higher vehicle mileage.

Figure 2.8 is sourced from the "China Automotive Low Carbon Action Plan 2020," illustrating how battery swapping effectively reduces wear and aging rates in batteries—thereby extending their lifespan and subsequently lowering carbon emissions per kilometer traveled. In a scenario where battery swap stations maintain a readiness rate of 10%, compared to gasoline vehicles, those utilizing swapped batteries can achieve a reduction in carbon emissions by approximately 34% after traveling 300,000 km and by about 36% after reaching distances of 450,000 km.

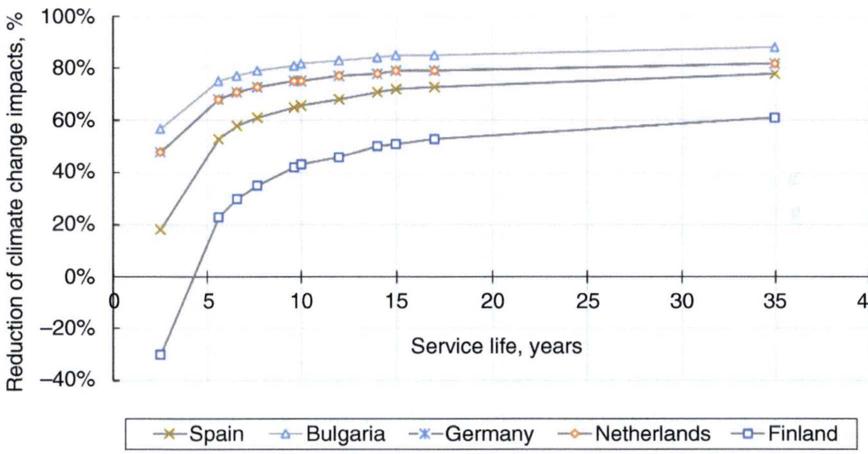

Figure 2.7 The impact of battery lifespan on greenhouse gas emissions.

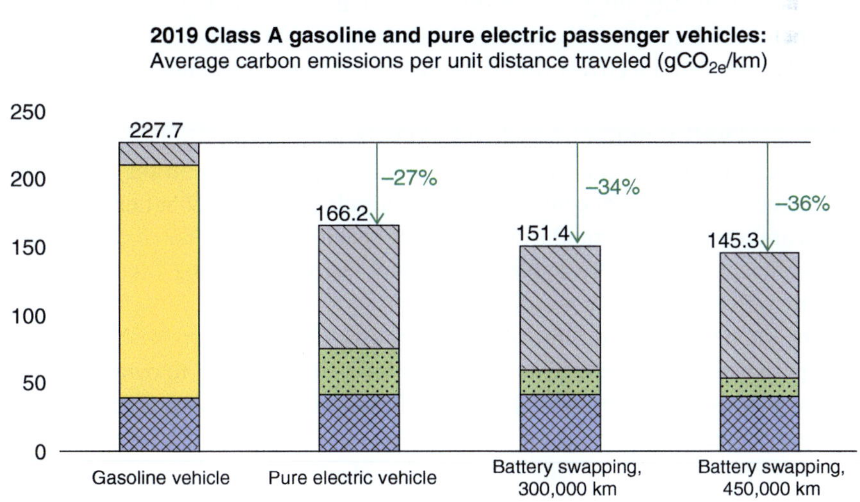

Figure 2.8 Estimation of carbon emission reductions through battery swapping.

2.3.3 Battery Swapping Promotes the Development of Standardized Batteries and Reduces GHG Emissions

The adoption of standardized battery design within the battery swapping model represents a pivotal initiative aimed at steering the EV industry toward a more environmentally sustainable future. Standardization not only contributes to reducing carbon emissions during the battery production process but also facilitates the optimization of manufacturing processes, thereby minimizing energy consumption and emissions.

In the electrode preparation phase, standardized design enhances and regulates critical steps in detail. By employing efficient and stable drying equipment and techniques, it is possible to control essential parameters such as temperature, humidity, and ventilation speed. Optimizing these drying conditions can significantly shorten drying times while decreasing energy usage, ultimately leading to lower levels of carbon emissions. During the pressing stage, controlling variables such as pressing pressure, temperature, and duration ensures that design specifications are met throughout the process. This consistency guarantees uniform structure and performance of electrode sheets, thereby improving production efficiency while reducing scrap rates and rework.

For commonly utilized lithium-ion batteries—characterized by their use of lithium carbonate along with polymer electrolytes and organic solvents containing lithium salts—the implementation of standardized design necessitates establishing unified electrolyte formulations and preparation protocols. Ensuring that each batch of electrolyte adheres to consistent standards reduces errors associated with variability or overuse. The utilization of efficient stirring and mixing equipment is crucial for achieving homogeneity in electrolyte composition; thus, standardized battery design mandates uniform stirring and mixing parameters which enhance both quality and consistency while mitigating waste resulting from uneven blending.

In conclusion, implementing standardized designs in lithium-ion battery electrolyte production can effectively diminish carbon emissions through strategies such as unifying formulas and procedures, optimizing equipment alongside processing conditions, as well as minimizing chemical usage.

To enhance the role of power batteries in achieving carbon neutrality, numerous manufacturers are now concentrating on the development of EV battery swapping systems and establishing standardized battery specifications. A notable example is CATL's "chocolate battery swap block" solution, introduced in 2022, which has significantly influenced the EV sector.

Firstly, the design of the "chocolate battery swap block" substantially improves user experience and cost-effectiveness. Users have the flexibility to combine different battery capacities according to their specific needs, thereby minimizing resource waste associated with purchasing incompatible capacities. For instance, users can select battery blocks with varying capacities based on distinct driving requirements, facilitating a more adaptable energy configuration.

Moreover, the open "multi-platform shareable" model presents broader developmental opportunities for the entire industry by fostering standardized and regulated growth. Through this multi-platform approach, various manufacturers can collaboratively utilize standardized battery designs. This collaboration not only reduces research and development costs but also accelerates product launch timelines while enhancing interoperability and compatibility among products.

By optimizing production processes, improving equipment utilization rates, and decreasing energy consumption during manufacturing, companies can achieve greater production efficiency. This optimization leads to improved resource utilization and a reduction in carbon emissions associated with battery production.

In the same year, CATL achieved remarkable advancements in heavy truck battery swapping and standardized design. Its groundbreaking MTB (Module to Bracket) technology was initially implemented in the heavy truck battery swap project of SPIC Qiyuan Green Power. This technological innovation opens up new possibilities within the electric heavy truck battery swapping sector. By utilizing MTB technology, modules are directly integrated into the vehicle's bracket/chassis, enhancing system volume utilization by 40%, reducing weight by 10%, and attaining a system energy density of 170 Wh/kg. Furthermore, the battery supports a flexible combination scheme ranging from 140 to 600 kWh, boasting a cycle life of 10,000 cycles or ten years, along with an innovative U-shaped cooling system that effectively addresses heat dissipation challenges.

As EV battery swapping and standardization continue to advance, the potential for power batteries to play an increasingly significant role in achieving carbon neutrality becomes ever more apparent, paving new avenues for sustainable industry development.

Standardized battery designs enhance disassembly efficiency and purification processes, thereby minimizing energy consumption as well as material waste during recycling operations. With an anticipated large-scale retirement of standardized batteries on the horizon, clear advantages will emerge in formulating effective recycling strategies as well as optimizing production line planning.

As the battery swapping model gains widespread adoption in the future, it will become feasible to consider recyclability from the initial design phase of batteries. Currently, researchers suggest that standardized battery designs could incorporate solid conductive wires instead of flexible cables, thereby facilitating the easy disassembly of large battery cells. These cells can be conveniently separated into anode, cathode, and polymer separator layers. The methods employed for assembling battery components and selecting electrode materials will significantly influence recycling processes.

The use of water-based adhesives facilitates the separation of current collectors, and detachable packaging technologies aid in the recovery of electrode powders.[1] Traditional hydrometallurgical recycling often involves compromising the crystal structure of materials and converting metallic elements into ions within a solution, resulting in substantial consumption of chemicals and energy resources. Consequently, standardized batteries should aim to circumvent challenges associated with lithium-ion battery recycling.

---

[1] Thompson, D. L., Hartley, J. M., Lambert, S. M., Shiref, M., Harper, G. D., Kendrick, E., Anderson, P., Ryder, K. S., Gaines, L., Abbott, A. P. J. G. C., The importance of design in lithium ion battery recycling–a critical review. *Green Chemistry* 2020, 22 (22), 7585–7603.

3

Technological Innovation and Standardization Process of Battery Swapping

3.1 Literature Visualization Analysis and Technological Research Progress

3.1.1 Global Research Hotspots of Battery Swapping Technology

As a critical component within the energy replenishment infrastructure for new energy vehicles, battery swap technology demonstrates prominent interdisciplinary characteristics in its research focal points, accompanied by a synergistic co-evolutionary pattern encompassing technological innovation, policy frameworks, and economic considerations.

Figure 3.1 explicitly delineates the annual publication volume of battery swap technology-related papers indexed in the Web of Science database, providing a visual representation of the dynamic evolution trajectory from 1999 to 2023. A thorough analysis of the chart data reveals the following observations:

- **1999–2010 (Formative phase):** The publication output remained at relatively low levels with sluggish growth, indicative of limited academic attention during this period. This pattern likely corresponds to the preliminary technological exploration and foundational research stage in the field.
- **2011–2018 (Emergent development phase):** A steady upward trajectory in paper quantity emerged, signifying the gradual recognition of battery swap technology within academic circles. This phase reflects intensified research activities and growing scholarly interest.
- **2019–2024 (Accelerated growth phase):** The publication count exhibited exponential growth, reaching a local maximum in 2022 and maintaining robust expansion through 2023–2024. This dramatic surge substantiates the rapid technological advancement and widespread adoption of battery swap systems in recent years.

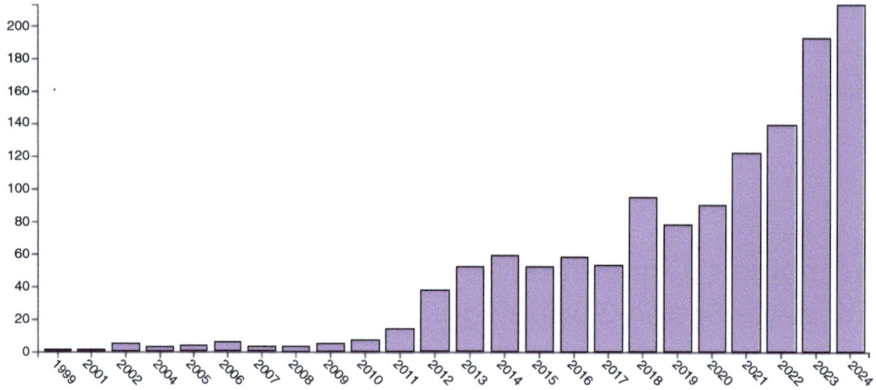

Figure 3.1 Web of Science annual publication output of battery swapping papers in the database.

The quantitative fluctuations in battery swap research publications stem from multifaceted determinants:

Industrial demand: The flourishing development of new energy vehicles (NEVs) has elevated the strategic importance of battery swapping as an efficient energy replenishment solution, which has stimulated vigorous research activities.

Policy drivers: Governmental initiatives worldwide promoting NEVs and associated infrastructure have strategically prioritized battery swap technology, significantly incentivizing academic and industrial research engagement.

Market dynamics: Intensifying competition among enterprises entering this sector has substantially amplified R&D investments, directly contributing to increased scholarly output.

Interdisciplinary synergy: The inherent multidisciplinary nature of battery swap technology—spanning mechanical engineering, electrical systems, and energy management—has attracted scholars from diverse disciplinary backgrounds, thereby accelerating knowledge production through cross-domain collaborations.

This analytical framework aligns with the tripartite co-evolution model of technological innovation, institutional mechanisms, and market forces, demonstrating how macro-level industrial transitions interact with micro-level research productivity in shaping academic landscapes.

Figure 3.2 further delineates the bibliometric distribution of publication types within the domain of battery swap technology. The chart explicitly demonstrates the overwhelming predominance of original research articles over other document categories, such as observational studies and review articles. This structural imbalance in scholarly output carries profound implications:

Methodological orientation: The dominance of research articles underscores the field's current emphasis on experimental operations and empirical analysis, reflecting an applied research paradigm aimed at addressing technical bottlenecks.

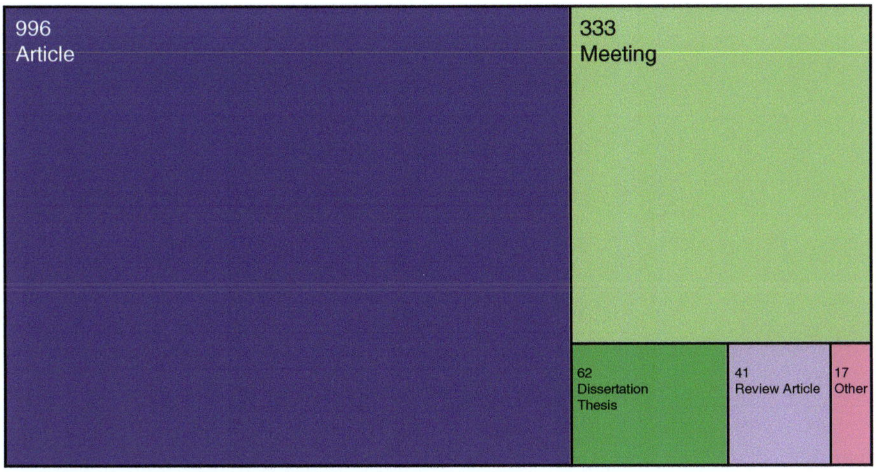

Figure 3.2 Distribution of types of battery swap papers in the Web of Science database.[1]

Theoretical development stage: The relative scarcity of review articles (constituting merely 2.9% of total publications) serves as a critical indicator of the field's developmental phase. Such paucity suggests that battery swap technology research remains in an accelerated growth period, where rapid knowledge accumulation has yet to coalesce into mature theoretical frameworks or systematic conceptual syntheses.

Globally, academic investigations into battery swap technology exhibit vigorous expansion, with research articles maintaining an 68.7% share of total publications from 2019 to 2024. This publication pattern not only validates the technology's status as an emerging research frontier but also mirrors the intense academic competition surrounding practical innovation.

3.1.1.1 Prospective Analysis

As the field approaches technological maturity thresholds, two evolutionary trajectories become discernible:

A projected surge in systematic reviews and meta-analyses is anticipated post-2025, driven by the necessity to consolidate fragmented empirical findings into unified theoretical constructs.

The eventual emergence of dedicated scholarly monographs will likely signify the transition from applied research dominance to theoretical systematization, potentially catalyzing paradigm shifts in energy replenishment methodologies.

This publication typology analysis aligns with Price's Law of scientific growth, wherein the proliferation of primary research precedes theoretical synthesis during disciplinary formation. The observed pattern further corroborates the triple-helix

[1] https://webofscience.clarivate.cn/wos/alldb/summary/b0517c78-3d74-4761-8e92-aaa9a7f57135-016b9363d0/relevance/1

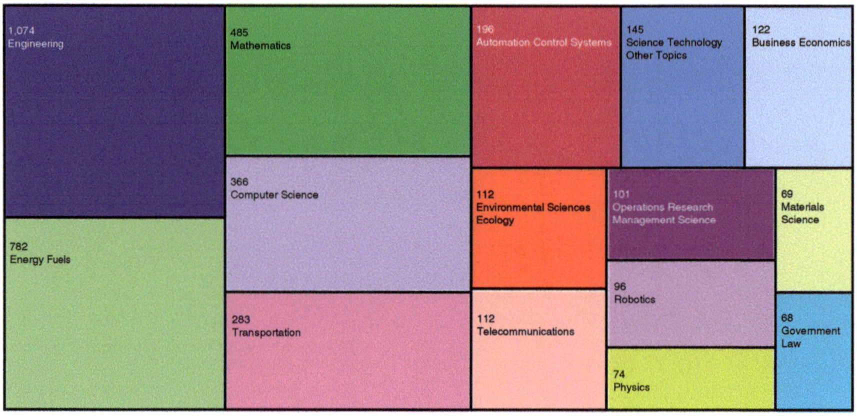

Figure 3.3 Classification of research fields for papers in the Web of Science database.[2]

model of innovation, where academic output dynamics directly reflect industry maturation levels and policy prioritization.

Figure 3.3 systematically illustrates the disciplinary distribution of battery swap technology research based on publication data, with the following key observations:

Core Disciplinary Contributions

Engineering dominates the field with 1,074 publications, reflecting its central role in addressing critical technical challenges such as system design, mechanical implementation, and operational optimization of battery swap infrastructure.

Energy and fuels rank second with 782 publications, highlighting extensive investigations into energy storage solutions, battery performance enhancement, and compatibility with renewable energy systems.

Mathematics contributes 485 publications, primarily focusing on algorithmic optimization, stochastic modeling for resource allocation, and predictive maintenance frameworks.

Computer Science accounts for 366 publications, emphasizing its growing influence in data-driven decision-making, intelligent control systems, and digital twin applications.

Supplementary Disciplinary Engagement

Other fields demonstrate targeted yet limited participation:

Automation and Control Systems addresses robotic automation in battery handling.
Business Economics explores cost-efficiency models and market scalability.
Environmental Science and Ecology investigates lifecycle environmental impacts.
Materials Science focuses on battery material innovations.

[2] "Analysis Results." Web of Science. 2023. https://www.webofscience.com

Additional disciplines, including operations research, telecommunications, physics, and government law, collectively contribute to niche aspects of technology development.

Interdisciplinary Synergy

The distribution underscores three fundamental characteristics:

Technical primacy: Engineering disciplines form the methodological backbone (64.7% of listed core publications).
Computational integration: Mathematics and Computer Science jointly enable advanced analytical capabilities (30.2% of core publications).
Cross-domain extension: Noncore fields collectively address socio-technical challenges in commercialization and sustainability.

Analytical Implications

The disciplinary architecture aligns with the technology's developmental trajectory, where engineering-driven innovation precedes systemic integration of computational and sustainability considerations.

The limited theoretical synthesis (evidenced by scarce review articles in Figure 3.2) correlates with this predominantly applied research focus across disciplines.

Future advancements will likely require deeper convergence between material science (battery chemistry), environmental engineering (circular economy), and policy studies (standardization frameworks).

Figure 3.4 shows the geographic distribution of contributions to battery swap technology research by country and region, based on Web of Science-indexed publications. The chart distinctly highlights China's overwhelming dominance in this

Figure 3.4 The distribution of battery swap papers by country of origin in the Web of Science database.[3]

[3] "Analysis Results." Web of Science. 2025. https://www.webofscience.com

field, with 1,062 publications—accounting for nearly 70% of the total mapped outputs. The United States follows distantly in second place with 220 publications, demonstrating its enduring influence in automotive innovation. India claims the third position with 94 publications, signaling its emergence as a significant contributor to sustainable mobility research.

Other notable contributors include the United Kingdom (59 publications), Australia (52), and Taiwan, China (42), though potential data inconsistencies require scrutiny (e.g., duplicate US and UK entries). Germany (40), South Korea (34), Canada (27), Japan (24), Italy (28), and Iran (22) exhibit specialized research focuses, ranging from industrial automation to cold-climate applications. However, these collective outputs from secondary contributors constitute less than 15% of total publications, underscoring fragmented global participation.

The distribution reveals extreme geographic concentration: the top three contributors (China, United States, India) collectively produce 88.5% of publications, while the remaining 10 listed countries/regions contribute only 11.5%. This stark disparity reflects asymmetric research capabilities shaped by industrial policies, technological infrastructure, and funding priorities. China's supremacy correlates with its national strategy to standardize battery swap infrastructure, whereas the United States maintains strength through private-sector automotive R&D ecosystems.

While the visualization effectively quantifies publication volumes, its interpretation necessitates caution. Apparent duplicates (e.g., conflicting US and UK. counts) suggest potential database categorization errors or institutional affiliation misclassifications. Researchers should cross-validate these findings with normalized metrics (e.g., publications per GDP/capita) to mitigate geographic bias.

The 10:1 publication ratio between China and the United States epitomizes divergent innovation paradigms—China's scale-driven, policy-backed approach versus the US's niche technology focus. Meanwhile, India's rising trajectory illustrates developing economies' strategic prioritization of modular energy solutions. The limited EU collective output despite individual country entries underscores challenges in regional research coordination.

This geographic concentration raises critical questions about knowledge democratization in sustainable transportation. While China's leadership accelerates technological maturation, broader global adoption requires intentional efforts to bridge the North–South research divide through open innovation platforms and standardized data-sharing frameworks.

3.1.2 Distribution of Core Journal Paper Topics over the Past 5 Years

By analyzing and summarizing the literature from the Web of Science database over the past five years (2021–2025), a total of 15 significant technical branches in the battery swap technology field were identified, as follows:

3.1.2.1 Modular Battery Design

Modular battery systems based on compartmentalized standardized architecture achieve rapid cross-model adaptation and hot-swap operations through physical interface decoupling and topological reconfiguration technologies. A typical example is Contemporary Amperex Technology Co. Limited's (CATL) Cell to Pack (CTP) technology, which significantly improves volume utilization by reducing the module level. This design paradigm can reduce the platform development costs of vehicle manufacturers by more than 30% and shorten the battery swap operation time to less than 3 minutes through mechanical decoupling design, in compliance with the energy replenishment efficiency requirements of the SAE J3105 standard for electric vehicles (EVs).

1. **Battery Standardization Protocol System**
   A cross-brand compatible standard covering mechanical interfaces, electrical parameters, and communication protocols is established, focusing on solving the uniformity issues of battery pack shape dimensions (such as the 200 × 1500 × 140 mm standard module of EVOGO's chocolate battery), high-voltage connector insertion and extraction force (ISO 17409 standard), and CAN bus data interaction protocols. This system can promote the formation of economies of scale in the industry. According to estimates by Boston Consulting Group, standardization can reduce the construction costs of battery swap networks by 22–35%.
2. **High-precision Positioning and Calibration System**
   Based on multi-sensor fusion technology (lidar + machine vision + inertial navigation), a vehicle pose recognition model at the millimeter level is constructed, and an improved iterative closest point (ICP) algorithm is used to achieve positioning accuracy of ± 0.5 mm. This system can ensure precise trajectory planning of six-degree-of-freedom robotic arms under complex working conditions, avoiding equipment collision risks caused by positioning deviations, in accordance with the technical specifications for the safe operation of industrial robots in ISO 10218-2.
3. **Battery Thermal Management Synergistic Control**
   An integrated composite thermal control system of phase-change materials and liquid cooling cycles is developed, and computational fluid dynamics (CFD) simulation is used to optimize the flow path design to ensure that the temperature gradient of the battery during the battery swap process is ≤ 3 °C. Experimental data show that this technology can increase the battery cycle life to more than 3,000 times (for the NCM811 system), and reduce the probability of thermal runaway propagation by 87% through active thermal runaway suppression algorithms.
4. **Rapid Connector Reliability Design**
   A multi-contact redundant layout and self-cleaning plug-in structure are adopted to achieve a contact resistance fluctuation rate of less than 5% under a plug-in and extraction life of 5,000 times (IEC 62860 standard). The liquid cooling-electricity composite interface technology displayed in Tesla's patent US20210021073A1 can maintain a temperature rise of less than 15 K and an energy loss rate of less than 1.8% during 150 kW power transmission.

3.1.2.2 Optimization of Battery Swap Station Equipment

5. **Fully Automated Battery Swap Robot System**
   A multi-robot arm collaborative control platform based on the Robot Operating System (ROS) architecture is developed, integrating hybrid force-position control algorithms and digital twin simulation verification systems. NIO's third-generation battery swap station achieves a 140-second battery swap cycle through the coordinated movement of 12 servo motors, improving operational efficiency by 40% compared to traditional solutions, and breaking through the daily average service capacity of 400 vehicles per station.
6. **Dynamic Inventory Optimization Model**
   A demand forecasting model based on long short-term memory (LSTM) neural networks is constructed, coupled with battery health status (State of Health, SOH) and time-of-use electricity price data, to establish a multi-objective optimization function for solving the optimal charging strategy. Real measurement data show that this model can increase the battery turnover rate to 85% and reduce the proportion of idle batteries from the industry average of 15–6%.
7. **Integrated Energy Management of Battery Swap and Storage**
   A vehicle-to-grid (V2G) bidirectional converter and energy router are developed to achieve the dynamic response capability of battery swap stations as nodes of a virtual power plant (VPP). The State Grid's demonstration project shows that a single station participating in frequency regulation ancillary services can create an annual revenue of about 480,000 yuan, while reducing the peak-valley load difference by more than 30%.
8. **Battery Full-life-cycle Assessment System**
   An electrochemical impedance spectroscopy (EIS) and deep reinforcement learning algorithm are integrated to construct a capacity fade rate prediction model (mean absolute error $< 1.5\%$). This system supports the establishment of a graded residual value assessment standard, providing technical endorsement for the asset securitization of power batteries.
9. **Mobile Battery Swap Equipment Integration Technology**
   A modular container design is adopted, integrating emergency power supplies and automatic leveling systems to achieve a rapid deployment capability of 72 hours. The mining area battery swap solution verified by XCMG Group shows that this technology can increase the operation efficiency of mining trucks by 25%, with a diesel substitution rate of more than 90%.

3.1.2.3 Ecological Construction Technology

10. **Battery Bank**
    The Battery Bank leverages digital twin and blockchain technologies to achieve full-life-cycle management and cross-platform circulation of battery assets, supporting a new type of energy service model characterized by "pay-as-you-use." Practices by CATL and NIO have shown that this technology can reduce the initial vehicle purchase cost for users by 40%, while increasing battery utilization to over 85%.

11. **BaaS (Battery as a Service) Service Architecture**
    Based on blockchain smart contracts for battery asset rights confirmation, and employing zero-knowledge proof technology to protect user privacy, this architecture enables cross-operator settlement and redemption. The practice case of BAIC BluePark has demonstrated that the BaaS model can reduce user acquisition costs by 40%, with an annual growth rate of battery asset management scale reaching 300%.
12. **Battery Swap Network Scheduling Algorithm**
    A spatiotemporal coupled queuing theory model is constructed, integrating real-time traffic flow data (GIS-TMC) and user behavior profiles to develop a dynamic route planning algorithm. During the Hangzhou Asian Games, the application of this algorithm reduced user waiting time during peak hours by 58%.
13. **Second-life Battery Sorting Technology**
    A multi-dimensional retired battery evaluation system is established (capacity, internal resistance, self-discharge rate), and an automatic sorting production line based on machine vision is developed. The empirical project by BYD has shown that the levelized cost of storage (LCOS) for reassembled battery packs in energy storage scenarios can be reduced to 0.35 yuan/Wh, a 60% decrease compared to new batteries.
14. **Energy Internet Data Protocol**
    The ISO 15118 standard is expanded to establish a cross-platform interaction framework covering SOC (State of Charge), SOP (State of Power), and SOE (State of Energy) parameters. This protocol supports vehicle-station-grid collaborative optimization. A pilot project in Germany has verified that it can enhance the renewable energy accommodation capacity of the regional power grid by 12%.
15. **Multi-level Security Protection System**
    A three-dimensional defense architecture is constructed, comprising physical protection (explosion-proof and pressure-relief enclosures), cybersecurity (national cryptography algorithm encryption), and risk hedging (battery insurance securitization). The Bowtie model is used for risk quantification analysis, reducing the system failure probability (PFD) to the order of 10^{-6}.

3.1.3 Research Progress and Future Technological Directions

The future development of battery swap technology will feature a multi-dimensional innovation landscape, primarily manifested in three core areas:

3.1.3.1 Technical Foundation and Energy Synergy

Battery Standardization and Sharing Technology

Battery standardization and sharing technology will serve as the cornerstone of the industry. By unifying parameters such as the voltage and dimensions of battery packs, cross-brand compatibility can be achieved. This, combined with a sharing and leasing model, will enhance the utilization rate of resources.

Intelligent Upgrades

Intelligent upgrades will propel battery swap stations to transition toward AI-driven operations. Big data will be utilized to forecast user demand, and automated robotic arms will facilitate battery swaps within minutes.

Deep Integration with the Power Grid

The deep integration with the power grid will be exemplified by battery to grid (B2G) technology and microgrid integration. Batteries will function as energy storage units for the power grid, stabilizing the fluctuations of renewable energy sources and enabling dynamic charging and discharging during off-peak and peak electricity price periods.

Full-life-cycle Management

Full-life-cycle management will extend the value chain of batteries through second-life applications. Combined with real-time health monitoring to optimize charging and discharging strategies, it will form a closed-loop system of "vehicle use—energy storage—recycling."

3.1.3.2 Operational Optimization and User Experience

Dynamic Operation System

The dynamic operation system will integrate multi-power charging configurations and real-time scheduling algorithms to smartly allocate resources based on grid load and user demand.

User Behavior-based Predictive Models

Predictive models based on user behavior analysis will accurately anticipate battery swap needs and, in conjunction with personalized recommendation systems, provide optimal service solutions.

Battery Swap Station Layout

The layout of battery swap stations will move beyond the single-node model. Through optimized site selection and scale, as well as coordinated layout with charging stations, a grid-based energy service network will be constructed. This network will take into account traffic flow, grid capacity, and user density to maximize efficiency.

3.1.3.3 Safety System and Economic Model

Safety Protection System

The safety protection system will build dual safeguards through thermal runaway early warning mechanisms and reliable contact piece design. Smart monitoring technology will be employed to achieve full visibility of battery status.

Commercial Innovation

Commercial innovation will be reflected in the combination of differentiated pricing mechanisms and cost-benefit models. An elastic billing system based on battery charge and usage scenarios will be established, while optimizing the full life cycle costs, such as equipment depreciation and maintenance.

Multi-objective Optimization Algorithms

Multi-objective optimization algorithms will balance key indicators such as operating costs, waiting time, and battery life. They will drive battery swap stations to transform from single service points to smart energy hubs.

3.2 Patent Layout and Core Technology Competitiveness

3.2.1 Global Trend in Battery Swap Patent Applications

Figure 3.5 presents that between 2010 and 2024, there was a remarkable increase in China's patent applications related to battery swapping, with the most significant growth occurring in 2016 when the number of patent applications rose by 183 from the previous year to reach 407. By 2023, the number of patent applications in this sector had soared to 3,677, marking the highest within this timeframe. This upward trend correlates closely with the rapid advancement of China's NEV industry and substantial policy backing.

Japan, a traditional powerhouse in automotive technology, has also experienced steady growth in patent applications for battery swapping. In 2016, Japan witnessed an addition of 84 patents compared to 2015, bringing the total to 271, which was the largest annual increase during this period. By 2023, Japan's patent applications in this field had reached 573, indicating the highest number of patents and reflecting its ongoing investment in innovation during the automotive electrification transition.

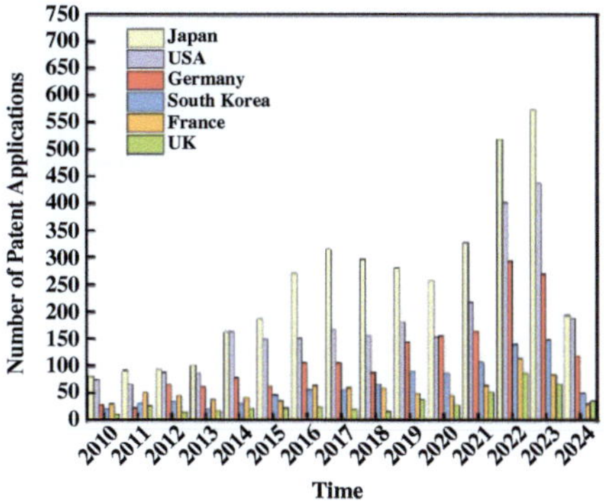

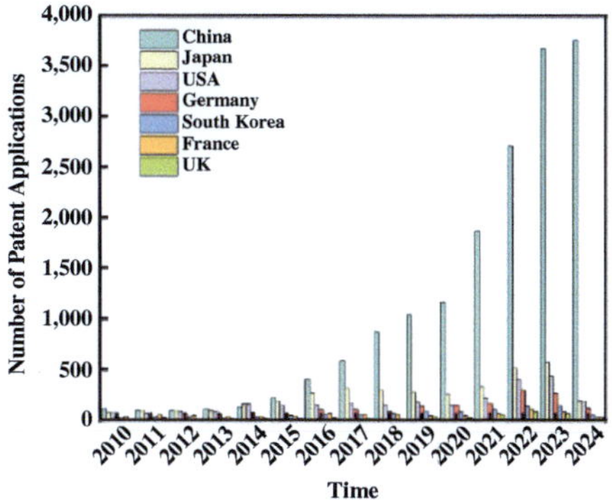

Figure 3.5 Annual patent application volume for battery swapping in the seven major countries.

The United States has shown an increasing trend in patent applications in this sector. In 2022, there was a growth of 186 patents compared to 2021, reaching a total of 403, representing the largest annual increase in the United States. In 2023, the number of US patent applications further rose to 438, the highest within the period, demonstrating vibrant activities in emerging automotive technologies.

Germany, despite having a relatively lower number of patent applications in battery swapping, has shown a certain growth trend. In 2022, the country saw an addition of 128 patents compared to 2021, reaching 293, which was both the largest annual increase and the highest number of patents for Germany.

South Korea experienced an addition of 34 patents in 2022 compared to 2021, reaching 141, marking its largest annual increase. In 2023, the number of patent applications in South Korea further rose to 148, the highest within the period.

France added 49 patents in 2022 compared to 2021, reaching 114, representing both the largest annual increase and the highest number of patents.

The United Kingdom saw an addition of 35 patents in 2022 compared to 2021, reaching 87, which was the largest annual increase and the highest number of patents.

Overall, the global growth in patent applications for battery swapping indicates increasing innovation in NEV infrastructure. This not only offers technological support for promoting automotive electrification and sustainable development but also signals intensifying competition among nations in establishing battery swapping technology standards and market presence. It holds significant implications for advancing the technology and shaping the market landscape of the global NEV industry.

Figure 3.6 intuitively presents the global distribution and proportion of patent offices for battery swapping patents via a world map. The map uses different colors and arrows to indicate the proportion of battery swapping patent applications received by each country and region. China leads with 49.31% of global battery swapping patent applications, closely linked to its fast-growing NEV industry and

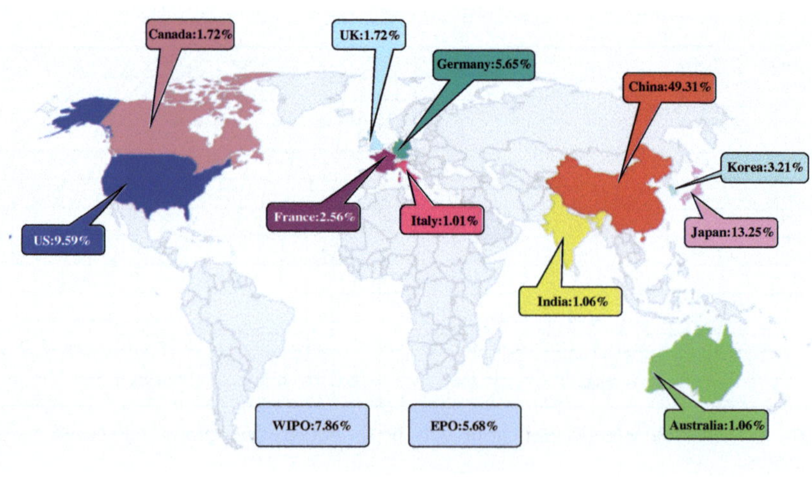

Figure 3.6 Ranking of patent counts by receiving offices.

supportive policies. The United States follows with 9.59%, reflecting its active R&D and patent applications in this field. Japan and Korea rank third and fourth with 13.25% and 3.21% respectively, showing their positive patent layout in this field.

France, Germany, Italy, Canada, the United Kingdom, India, and Australia also hold certain shares of global battery swapping patent applications. Most of these countries and regions are developed economies with advantages in the NEV industry. The map also shows that the WIPO and EPO account for 7.86% and 5.68% of patent applications, highlighting their key role as international patent offices.

Overall, the map reveals that battery swapping technology patents are concentrated in a few countries and regions, with China, the United States, Japan, and Korea being the major players. This reflects the current regional distribution of global battery swapping technology development and suggests fiercer future competition in this field. As a crucial charging method for NEVs, battery swapping technology's innovation and patent layout will significantly impact industrial progress and market structure. Countries need to boost R&D investment and international cooperation to advance battery swapping technology and ensure sustainable industrial development.

As shown in Figure 3.7, the patent applications for battery swap technology in China demonstrate distinct phase characteristics:

Technological Incubation Period (2006–2015)

During this period, the annual number of patent applications remained at a two-digit level, with a compound annual growth rate of less than 15%. The characteristics

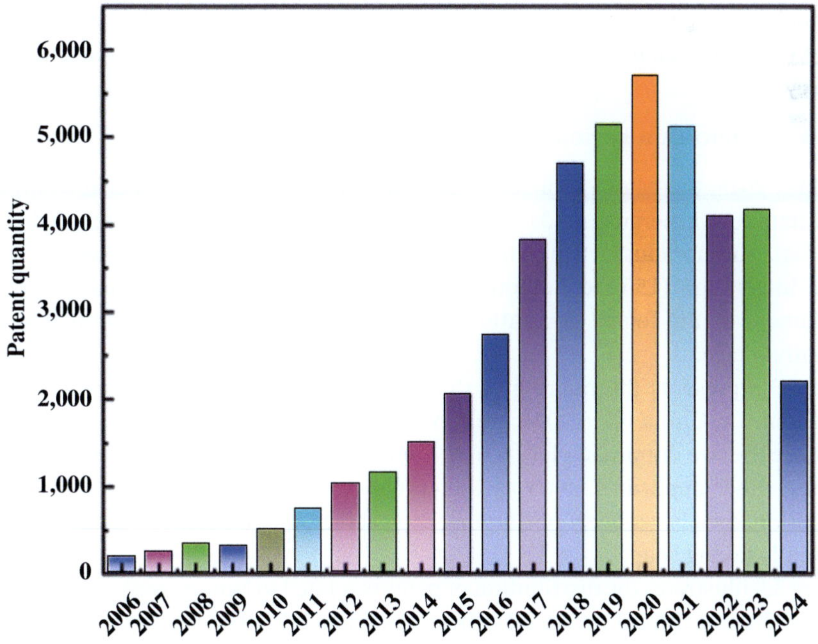

Figure 3.7 Trend of patent applications for vehicle battery swapping in China in 2024.

of the early stage of industry development were evident, mainly manifested in the unclear technological routes, limited participation of market entities, and concentrated R&D investment in the exploration of basic structures. The main patent applicants during this period were research institutes, and the proportion of industry application-oriented patents was less than 30%.

Explosive Growth Period (2016–2021)

The number of patent applications grew rapidly at an average annual rate of 78%, reaching a peak in 2020. This surge was strongly associated with three factors: First, the penetration rate of NEVs exceeded the critical threshold of 5%, giving rise to the demand for diversified energy replenishment technologies. Second, the joint issuance of the "Development Plan for the New Energy Vehicle Industry (2021–2035)" by the Ministry of Industry and Information Technology and three other ministries provided policy impetus. Third, the acceleration of the standardization process of power batteries enhanced the commercial feasibility of battery swap technology. During this period, the proportion of enterprise applications increased to 82%, and the proportion of utility model patents reached 65%.

Rational Development Period (2022–2024)

The growth rate of patent applications was adjusted to an average annual rate of 22%, but the absolute increment remained high. The technological evolution during this period was characterized by two main features: First, the patent quality index (PQI) increased from 1.2 to 1.8, and the proportion of invention patents exceeded 40%. Second, the technological direction showed clear differentiation, with battery quick-locking mechanisms (accounting for 31% of patents), thermal management systems (24%), and cloud-based battery management systems (18%) emerging as the three key areas.

Frontier Exploration Period (2025)

As of May 2025, the patent application data already showed new trends in technological iteration. Emerging fields such as solid-state battery adaptation technology (27 cases), autonomous driving shuttle systems (19 cases), and modular energy storage integration (15 cases) accounted for 43% of the total, indicating that the development of the fourth generation of battery swap technology has entered a substantive stage.

Figure 3.8 presents the ranking of the main applicants for automotive battery swap patents in China up to 2024. Aulton New Energy Technology and Kabushiki Kaisha are the two dominant players in the field of battery swap technology patents. However, other applicants, such as General Motors, Qualcomm, and State Grid Corporation, are also actively positioning themselves, and a competitive landscape is gradually taking shape.

Figure 3.9 illustrates the distribution of patents related to battery swap technology across different technical theme classifications up to 2024. The horizontal axis represents technical effect phrases, while the vertical axis represents technical theme classifications. The size of each bubble indicates the number of patents.

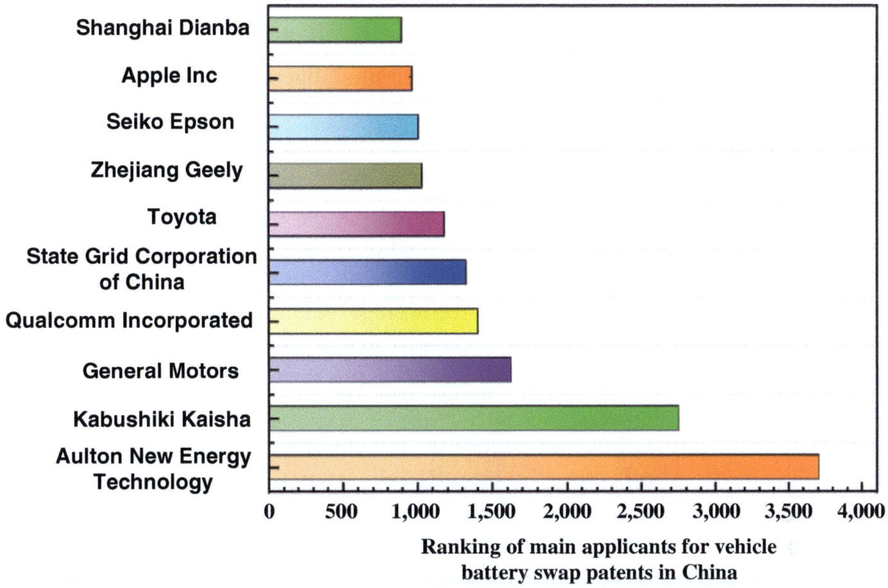

Figure 3.8 Ranking of main applicants for vehicle battery swap patents in China.

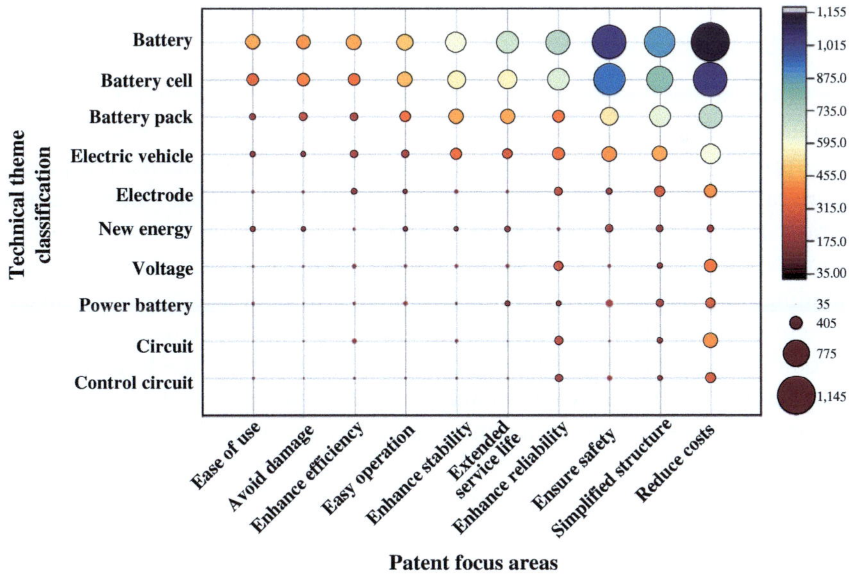

Figure 3.9 Patent focus areas.

Among the technical theme classifications, patents related to "batteries" are the most numerous, especially under the technical effect phrase "low cost," where the number of patents reaches 1,153. This indicates that battery technology is a key research focus in terms of cost reduction. Patents related to "battery cells" and "battery packs" are also substantial, with 1,026 and 711 patents, respectively, under the "low cost" effect, showing that these technologies also emphasize cost control. Additionally, themes such as "EVs," "electric motors," "new energy," "voltage," "power batteries," "circuits," and "control circuits" have a significant number of patents under the "low cost" effect, highlighting the close attention to cost control across the entire battery swap technology field.

Regarding technical effect phrases, "low cost" is the direction with the most patents, indicating that reducing the cost of battery swap technology is a core focus of current patent applications. Meanwhile, "improving safety" also has a considerable number of patents under the "batteries" theme, showing that safety is another important direction for battery swap technology patent applications. Moreover, "improving reliability" has numerous patents across multiple themes, indicating that reliability enhancement is a key aspect of the development of battery swap technology.

Overall, patent applications for battery swap technology cover multiple technical theme classifications, reflecting the interdisciplinary nature of the field. Cost control, safety, and reliability are the main focuses of current patent layouts in battery swap technology. In the future, as technology continues to advance, research and development in battery swap technology is likely to continue focusing on these key areas, with more innovations and patent applications expected in improving efficiency, extending lifespan, and other directions.

3.2.2 Patent Comparisons of Key Enterprises

Figure 3.10 illustrates the patent distribution of NIO's Battery Swapping Technology in the automotive field. As shown in the figure, the horizontal axis represents different enterprises, while the vertical axis indicates the number of patents. The patent types are categorized into four kinds: invention patent publications, invention patents granted, utility models, and design patents.

Taking NIO Co., Ltd. as an example, its performance in patents is relatively outstanding. The number of utility model patents reaches 212, with 202 invention patent publications, 36 design patents, and 15 invention patents granted. The proportion of utility model patents in NIO's patent portfolio is the largest, which reflects that NIO is quite active in the improvement and innovation of automotive technology. It focuses on optimizing the practical aspects of product structure and functions to enhance the performance and user experience of automobiles. The number of invention patent publications ranks second, indicating that NIO has also made certain investments in core technology research and development, actively applying for invention patents to protect its technological innovation achievements. Although the number of invention patents granted is relatively small, it still reflects the quality and value of its technological innovation to some extent. With 36 design patents, it shows that NIO attaches importance to the exterior styling and design aesthetics of automobiles. It is committed to creating distinctive and attractive product appearances to meet consumers' aesthetic needs for product appearance.

Compared with NIO Co., Ltd., other enterprises, such as Shanghai NIO Co., Ltd., have fewer patents overall, and the distribution of different patent types also varies.

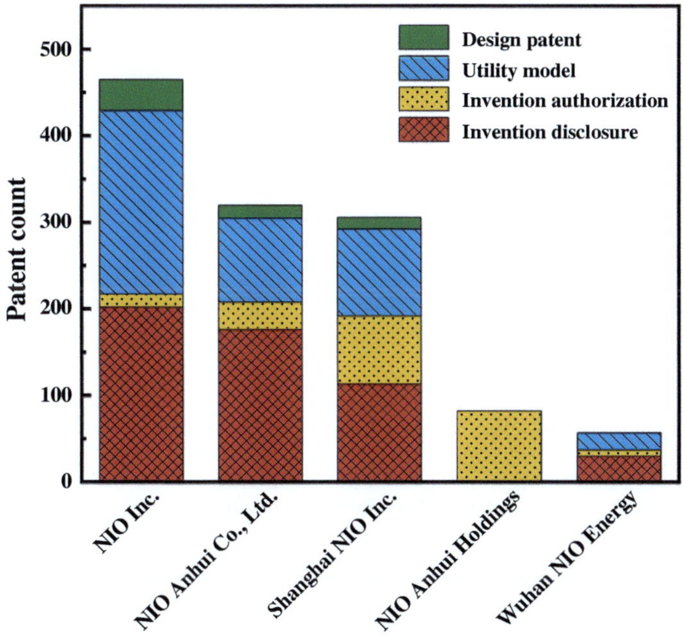

Figure 3.10 Patent distribution of NIO's battery swapping technology.

This difference may be related to the business focus, development stage, and innovation capacity of each enterprise. The patent advantage of NIO Co., Ltd. may be attributed to its technological accumulation in the field of NEVs, continuous research and development investment, and keen insight into market demand. This enables it to achieve rich patent results in multiple technical links and product design, thus occupying a favorable position in the fierce market competition.

Overall, the chart reveals the different strategies and focuses of automotive enterprises in patent layout. NIO Co., Ltd., with its accumulation of utility model and invention patents, demonstrates strong innovation ability and technical competitiveness in terms of patent quantity and type distribution. This provides strong technical support and guarantees for the sustainable development and market expansion of the enterprise, and also plays a positive exemplary role in the technological innovation and product upgrading of the entire automotive industry. Analysis of patent data can provide important references for automotive enterprises to optimize their research and development directions, strengthen intellectual property protection, and formulate market competition strategies, which is conducive to promoting the healthy and sustainable development of the automotive industry.

Figure 3.11 presents the distribution of patent quantities and types among selected enterprises in the automotive field. The patent types are categorized into invention publications (red), invention grants (yellow), and design patents (blue).

The chart illustrates that Tesla has over 70 patents, comprising 42 invention publications, 23 design patents, and 11 invention grants. The substantial number of invention publications reflects Tesla's strong innovative capacity and technological foresight.

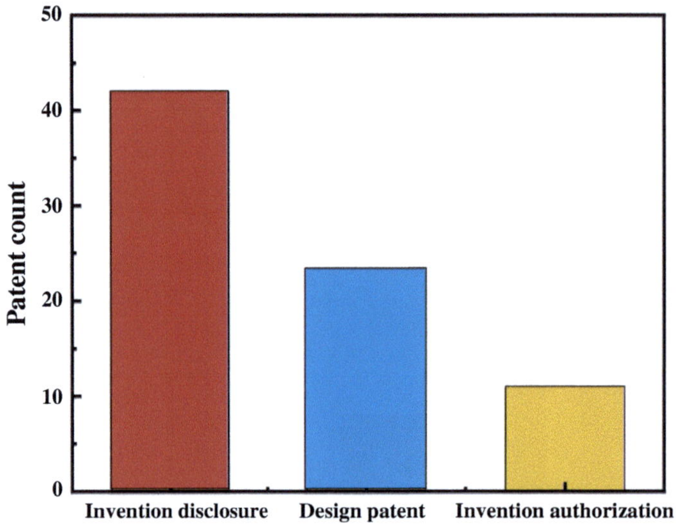

Figure 3.11 Patent distribution of Tesla's battery swapping technology.

By actively disclosing its research and development outcomes, Tesla enhances its technological influence within the industry and seeks legal protection for its innovations. The 23 design patents indicate that Tesla places great emphasis on product appearance, striving to enhance brand recognition and product appeal through a unique design style. This meets consumers' aesthetic demands and pursuit of personalization in automotive appearance. Additionally, the 11 invention grants demonstrate the tangible achievements of Tesla's technological innovation process. These granted patents serve as powerful tools in market competition, effectively preventing infringement by competitors and consolidating Tesla's leading technological position.

Overall, Tesla's lead in patent quantity and type distribution demonstrates its innovative advantages and strong technological competitiveness in the automotive industry. Its preeminence in multiple dimensions, including invention publications, design patents, and invention grants, not only provides solid technical support and legal protection for its own development but also exerts significant influence on the technological innovation and product design trends in the entire automotive industry. This patent advantage likely stems from Tesla's long-term and sustained research and development investment in the field of NEVs, its excellent engineering team, and its precise grasp of market demands and future technological trends. In comparison, the deficiencies in patent layout of other companies may limit their competitiveness and technological innovation capabilities in the market, providing directions for improvement in their future research and development investments and innovation strategies.

Figure 3.12 illustrates the distribution of patent quantities and types among Aulton New Energy Automotive Technology Co., Ltd. and several other companies. The horizontal axis represents different companies, while the vertical axis indicates the number of patents. The patent types are categorized into invention publications (red), utility models (blue), invention grants (yellow), and design patents (green).

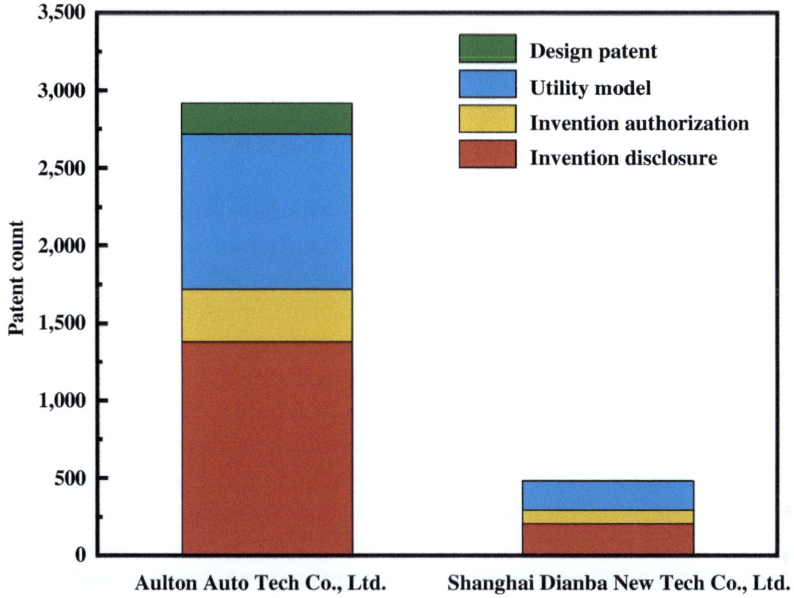

Figure 3.12 Patent distribution of Aulton New Energy's battery swapping technology.

Aulton New Energy Automotive Technology Co., Ltd. takes the lead in patent quantity with a total of nearly 3,000 patents. Among them, the number of utility model patents reaches 999, which is the largest proportion. This indicates that Aulton has made significant investments in the practicality and innovation of automotive technology, focusing on improving and optimizing existing technologies to enhance product performance and reliability. There are 1,383 invention publication patents, which is also a considerable number, reflecting Aulton's active engagement in technology research and development as well as innovation. It actively explores new technologies and timely applies for patents to protect its technological advantages. The number of invention grant patents is 335. Although it is relatively small compared with the number of invention publications, it is still a significant figure, reflecting the actual achievements and recognition obtained by Aulton in the process of technological innovation. There are 200 design patents, which shows that Aulton also pays attention to the appearance design of automobiles and is committed to creating attractive and recognizable product appearances to meet consumers' aesthetic needs for product appearance.

Other companies, such as Shanghai Dianba New Energy Technology Co., Ltd., have relatively fewer patents, and there are significant differences in the distribution of different types of patents. For example, Shanghai Dianba New Energy Technology Co., Ltd. has a certain number of patents in utility models and invention publications, but its overall scale is far behind that of Aulton New Energy Automotive Technology Co., Ltd.

In summary, Aulton New Energy Automotive Technology Co., Ltd. demonstrates strong innovative capacity and technological advantages in patent layout. Its comprehensive layout in multiple dimensions, including utility models, invention publications, invention grants, and design patents, provides solid technical support and legal protection for its development in the field of NEVs. This helps to consolidate its leading position in the industry. In comparison, other companies have a significant gap in patent quantity and type distribution with Aulton, which may imply that they are relatively weaker in research and development investment, innovation capacity, and market competitiveness.

Figure 3.13 illustrates the patent holdings and distribution of patent types in the automotive battery swapping field for the State Grid Corporation of China and its subsidiaries. Here, SGC stands for State Grid Corporation, SGCC is the abbreviation for State Grid Corporation of China, CEPRI Co., Ltd. represents China Electric Power Research Institute Co., Ltd., and CEPRI is the abbreviation for China Electric Power Research Institute Co., Ltd. It can be observed that the SGCC occupies a significant leading position in the field of automotive battery swapping patents. Its patent portfolio fully reflects its strong technological innovation capabilities and market competitiveness. According to the data, the total number of patents held by the SGCC, in the Chinese market is close to 21,000, far exceeding other companies in the industry. Among them, there are 7,420 published invention patents, 6,349 utility model patents, 4,207 granted invention patents, and 63 design patents. This distribution not only reflects the company's deep accumulation in basic technology research and development (invention patents) but also demonstrates its ability to rapidly implement technology (utility model patents). The high-quality granted invention patents further validate the practical value of its innovative achievements.

Although the subsidiaries' performance in the patent field is not as prominent as that of the parent company, they still exhibit a certain level of technological activity. For example, institutions such as the CEPRI and TEPC of the SGC have

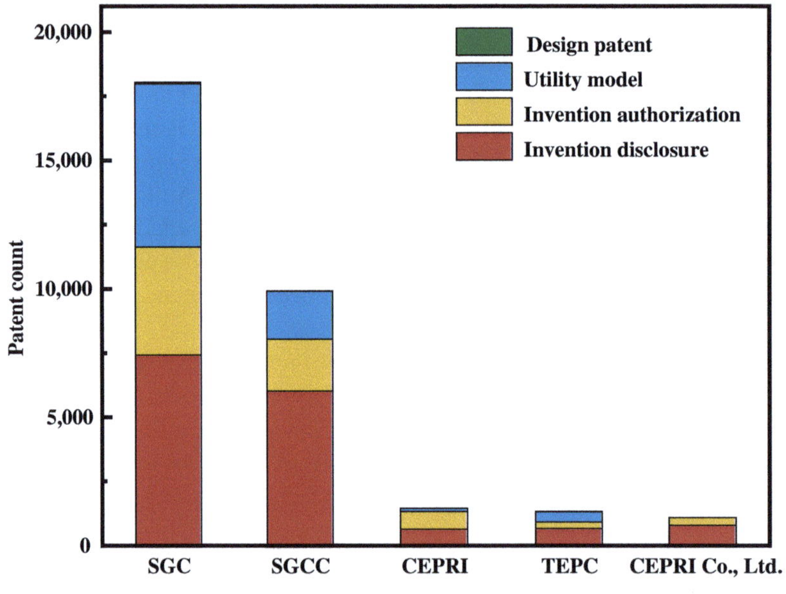

Figure 3.13 Patent distribution of State Grid's battery swapping technology.

submitted some patent applications in the relevant fields. However, both in terms of quantity and diversity of types, there is a significant gap compared with the State Grid Corporation of China. This indicates that the core research and development resources of the State Grid Corporation of China are highly concentrated, while the subsidiaries mainly undertake auxiliary technological exploration to support the overall strategic layout of the parent company.

From an industry impact perspective, the patent advantages of the State Grid Corporation of China not only provide legal protection for its technological development but also promote the technological standardization and industrialization process of the entire power industry. The comprehensive patent layout enables the company to maintain a leading position in both domestic and international markets, while setting a benchmark for industry innovation and further consolidating its leadership in the energy field.

Figure 3.14 illustrates the patent holdings of LG New Energy Co., Ltd. in the field of EV battery swapping, revealing that LG New Energy Co., Ltd. holds a significant advantage in terms of patent quantity, with a total of nearly 150 patents, which substantially exceeds that of other companies. The distribution of patent types indicates that a substantial proportion consists of invention disclosures, signifying that LG New Energy Co., Ltd. is highly active in the realms of technological research and development as well as innovation. The company is proactive in applying for invention patents to safeguard its technological innovations. Additionally, the number of invention grants is also considerable, reflecting the actual achievements and recognition garnered by the company throughout the process of technological innovation.

All data in this section are sourced from the Zhihuiya Patent website.

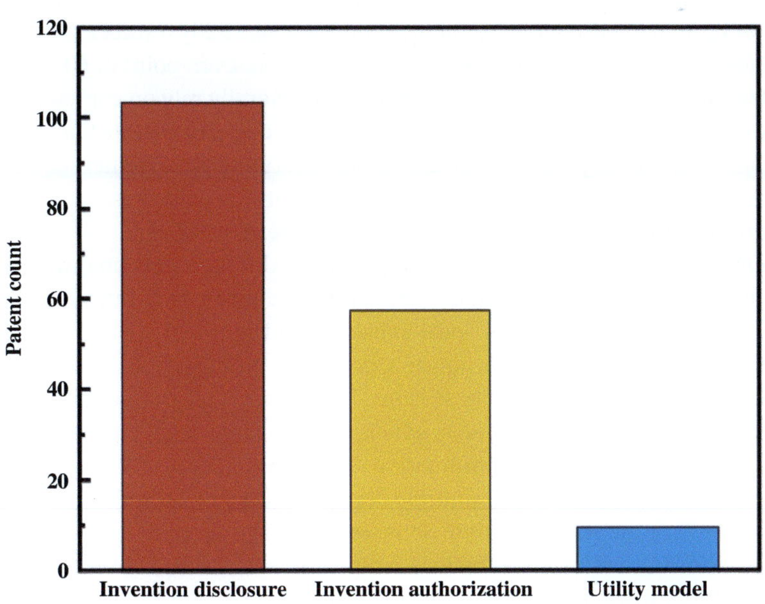

Figure 3.14 Patent distribution of LG New Energy Corporation's battery swapping technology.

3.2.3 Technological Bottlenecks in China's Battery Swapping Sector

The development of China's battery swap technology sector has been rapidly advancing under the dual impetus of policy support, technological innovation, and market demand. However, it still faces multiple development challenges, mainly reflected in the following aspects:

1. **Insufficient Standardization and Compatibility of Technology**
   Nonuniform battery specifications: Significant differences in battery specifications and interface protocols among various automakers and battery manufacturers make it difficult for battery swap stations to provide cross-brand services. For instance, although Contemporary Amperex Technology Co., Limited (CATL) has attempted to achieve modular compatibility for multiple vehicle models through its "chocolate battery swap block" design, the actual number of compatible models remains limited.

   Need for enhanced technological maturity: High-frequency plug-and-play of battery swap equipment may lead to interface wear and poor contact issues, while insufficient quality control of third-party batteries may increase safety risks (such as short circuits or explosions). Additionally, the intelligent dispatching systems and battery health management technologies of battery swap stations still need optimization to improve operational efficiency.
2. **High Infrastructure and Operational Costs**
   Substantial construction investment: The construction cost of a single battery swap station ranges from 1.5 to 5 million yuan (depending on the technological generation), and it also requires a reserve of batteries (14–30 batteries per station). For example, NIO's 3,117 battery swap stations have already exceeded 10 billion yuan in fixed asset investment, but the average daily number of swaps per station in 2024 is only 30 times, far below the break-even point of 60 times.

   Unclear profit model: Battery swap companies generally rely on capital infusion. Although the B-side market (such as taxi services) can establish a viable business model, the dispersed demand of C-side users and their higher requirements for convenience lead to significant operational cost pressures.
3. **Inadequate Policy Support and Regulatory System**
   Fragmented and difficult-to-implement policies: Although the country has included battery swapping in the "new infrastructure" and introduced a number of encouraging policies, local subsidies and national standards have not yet been fully unified. For instance, Chongqing provides subsidies of 400 yuan per kilowatt for battery swap stations, but cross-brand compatibility has only been improved from 30% to 80%, which restricts large-scale replication.

   Lack of inter-departmental coordination: The construction of battery swap stations involves approvals from multiple departments such as land use, power grids, and environmental protection. Some regions have simplified the process through "priority road rights + battery swap station rewards," but a national regulatory framework has not yet been established.

4. **Lagging User Acceptance and Market Education**
Cognitive bias and trust crisis: Users have doubts about the safety and battery quality of the battery swap model. Some car owners believe that the cost of battery swapping is higher than charging (for example, the cost of swapping a 75-degree battery is 30–50 yuan more expensive than fast charging), and the insufficient coverage density of battery swap stations affects the experience.
Competition from charging infrastructure: The advancement of fast charging technology (such as 800 V high-voltage platforms) and the increased coverage of charging networks have weakened the efficiency advantage of battery swapping. In 2024, the proportion of fast charging piles for new energy passenger vehicles has exceeded 50%, and users are more inclined to choose charging for energy replenishment.
5. **Challenges in Industry Chain Coordination and Ecosystem Construction**
Conflicts in interest distribution: Battery costs account for 35–40% of the total vehicle cost, and automakers are reluctant to give up battery profits, leading to slow progress in standardization. For example, automakers are concerned that the involvement of third parties may result in the loss of battery control rights, affecting their own benefits.
Complex battery asset management: The full-life-cycle management of batteries involves leasing, cascade utilization, and recycling, requiring the establishment of cross-industry collaboration mechanisms. For instance, CATL has attempted to address this issue through the "battery bank" model, but data tracking and regulatory rules still need to be improved.
6. **Controversies over Technical Adaptability in Specific Scenarios**
Divergent paths for commercial and passenger vehicles: Commercial vehicles (such as heavy trucks) are more suitable for the battery swap model due to their high-frequency operations and closed scenarios, while the passenger vehicle market faces a game of chess between charging and swapping. In 2024, the sales volume of battery swap heavy trucks accounted for 35% of the new energy heavy truck market, but the penetration rate of passenger vehicle battery swapping is less than 10%.
Risks in technology route selection: The battery swap model needs to balance standardization and flexibility. If the wrong technology route is chosen (such as the debate between chassis swapping and compartment swapping), it may lead to resource waste.

3.2.3.1 Future Breakthrough Directions

Accelerate the standardization process: The government leads in formulating unified battery specifications and battery swap interface standards to promote cross-brand compatibility (such as the Ministry of Industry and Information Technology's plan to release safety standards for commercial vehicle battery swapping in 2025).

Innovate business models: Explore a composite revenue model of "battery leasing + battery swap services + energy storage peak shaving" to improve the utilization rate of battery swap stations.

Policy and market synergy: Reduce corporate costs through subsidy inclination, land concessions, and carbon trading mechanisms, while also strengthening user education to improve market acceptance.

In summary, the development of China's battery swap technology sector needs to achieve breakthroughs in technical standards, cost control, policy coordination, and ecosystem construction in order to move from a "niche game" to large-scale application.

3.3 Domestic and International Standard Systems and Interoperability

3.3.1 Comparisons of Battery Swapping Standards for EVs

Standardization has long been a critical issue hindering the advancement of battery swapping technology. Key aspects include technical solutions for battery swapping, battery packs, and other components that are vital to the current development stage of the EV battery swapping industry. Standardization enhances system interoperability and ensures compatibility and interchangeability among batteries and battery swapping equipment from various manufacturers. This not only provides users with greater freedom of choice but also increases the accessibility of the battery swapping model.

Furthermore, standardization reduces both technology deployment and maintenance costs, yielding significant benefits for the construction and operation of battery swapping stations. Additionally, it plays a crucial role in ensuring the safety and reliability of battery swapping systems by establishing safety specifications and testing standards. The absence of standardization can lead to fragmentation within the battery swapping industry, ultimately diminishing its overall efficiency. Such fragmentation may result in different regions or enterprises adopting varying technologies and specifications, leading to incompatibility issues that restrict mobility and limit the operational range of EVs.

Moreover, standardization lowers market entry barriers, fosters competition, creates more opportunities for new enterprises to enter the market, and stimulates industrial innovation and development.

Although the revision of China's battery swapping standards has progressed significantly, the essence of standardization in battery swapping necessitates compromises among vehicle manufacturers, battery swapping operators, and between these two groups. Different stakeholders often possess varying technical preferences, complicating the process of reaching consensus and coordination. Vehicle manufacturers and battery swapping operators may prioritize their own interests, making it challenging to agree on technology and standards. Additionally, disparate local standards and group standards follow different technical pathways, resulting in a scenario where hardware interface sizes and communication protocols cannot be shared or interchanged effectively.

In recent years, battery swapping for EVs has garnered significant attention from the industry as a crucial technological pathway for energy replenishment, driven by robust market demand. In April 2021, the Ministry of Industry and Information Technology launched pilot projects to explore the application of battery swapping systems in NEVs, thereby further propelling the advancement of EV battery swapping technology. Major cities across the nation have been actively deploying and systematically promoting this initiative, enacting policies that support the implementation of battery swapping solutions.

Key enterprises within the industry have proactively expanded their market presence, establishing a relatively comprehensive industrial chain. Original equipment manufacturers (OEMs) such as NIO, BAIC New Energy, SAIC, Foton, CAMC, and SANY Group have accelerated their development of battery swappable models by introducing a series of innovative products. Battery manufacturers, including CATL, EVE Energy, Gotion High-Tech, and CAC, have developed products tailored for battery swappable vehicles to meet diverse operational needs. Additionally, operators of battery swapping stations such as Aulton, EVOGO, Enneagon Energy, GCL, and ZHILIWULIAN have rolled out service solutions that facilitate both vehicle operations and station functionality.

The fundamental requirement for battery swapping is the interchangeability among vehicles, battery swapping systems, and battery swapping stations. However, various brands currently adopt different technical approaches and solutions, which underscores the necessity for standardization to enhance interchangeability, safety, and product performance. In response to these needs, China has undertaken research on the standardization of EV battery swapping.

1. **Analysis of Battery Swapping Standard Requirements**
   Standards play a crucial role in regulating products, unifying requirements, and guiding technological advancement. The standardization of EV battery swapping has emerged as a significant factor in the industry's development. This necessitates a clear delineation of product solutions and technical pathways, the formulation of a comprehensive blueprint for the standard system, and the execution of research and standards development based on the urgency and importance of identified needs.

   In terms of safety, it primarily encompasses safety requirements during both the battery swapping process and subsequent usage. This includes considerations related to the installation, retention, and secure fixation of battery packs. Additionally, it addresses reliability concerns and safety risks associated with complex environmental factors such as vibration, temperature fluctuations, humidity levels, and contaminants that electrical connections, communication interfaces, and cooling systems may encounter during vehicle operation, as well as during battery swapping activities.

   In terms of interchangeability, unified structures and electrical parameters are essential for EV battery swapping. It is crucial to ensure the compatibility of battery swapping systems as well as their interfaces and mechanisms. Furthermore, attention must be given to the uniformity of the operational voltage range across battery swapping systems, coordination of charging and discharging performance, and compatibility of communication interfaces.

The heat management system within the battery swapping framework should also exhibit a degree of standardization, such as utilizing a common cooling medium.

Under the premise of ensuring EV battery swapping compatibility, it is imperative that facilities and equipment designed for executing battery swapping operations are standardized. This includes specifications regarding the dimensions of vehicles accommodated by battery swapping stations, connection interfaces between the battery swapping equipment and batteries themselves, operating space pathways and dimensions, sizes of battery pack racks within these stations, as well as communication protocols utilized by chargers.

All aforementioned aspects necessitate standardization through industry discussions, verification processes, and final determinations in order to establish a cohesive standardized plan that will guide the development of the EV battery swapping industry.

2. **Research on Chinese and Industry Standards**
To address the challenges of interchangeability, safety, and usability associated with EV battery swapping, China has engaged in extensive discussions and research focused on standardization. Progress in this area has been promoted through three key aspects:

First, there is a need to clarify the terminology and definitions related to battery swappable EVs. Currently, EVs are categorized into three types: pure EVs, hybrid EVs, and fuel cell EVs (Source: GB/T 19596-2017). According to existing definitions, a vehicle must be equipped with a battery system to qualify as a complete EV. Additionally, the tax system issues invoices for EVs during sales transactions. One significant approach to promoting battery swapping involves allowing consumers the option of purchasing vehicles without including the battery packs initially; this effectively lowers the upfront cost of acquiring an EV. However, current tax policies pose constraints on advancing the battery swapping model. Consequently, research within the EV sector has focused on refining terminology and definitions pertaining to battery swappable EVs—clarifying that vehicles lacking integrated battery systems can also be recognized as products within the broader category of EVs—thereby facilitating acknowledgment of optional battery sales by tax regulations.

Secondly, the advancement of battery swapping safety standards is crucial. Frequent battery swaps may introduce safety risks during the operation of battery-swappable EVs, necessitating in-depth research into safety standards for battery swapping within the automotive industry. In 2021, the industry established its first national standard for battery swapping safety, titled "Safety Requirements for Battery Swapping Electric Vehicles" (GB/T 40032-2021). This standard aims to mitigate safety risks associated with vehicle battery swapping and usage by imposing stringent requirements and testing protocols on both the entire vehicle and its components, thereby enhancing the overall safety of battery swappable vehicles. Simultaneously, the rapid development of commercial vehicles equipped with battery swapping capabilities—particularly heavy trucks—demands specific attention due to their unique rear-mounted design (with battery systems installed behind the cab) and more rigorous operating conditions compared to passenger vehicles that utilize underbody battery swapping designs. Consequently, there is a pressing need for tailored safety requirements regarding these designs. To address this gap, China has initiated research into establishing a new standard titled "Safety Requirements for Battery Swapping Electric Commercial Vehicles." This initiative proposes standardized criteria focusing on both design and operational safety to ensure sustainable growth within this sector.

Thirdly, it is essential to complete the research on the standardization scheme for battery swapping interchangeability. To facilitate seamless interchangeability among vehicles, battery swapping systems, and battery swapping stations, standardization must be implemented at both hardware and software levels. In terms of hardware, nine automotive industry standards have been established that delineate the mechanical dimensions, electrical interfaces, cooling interfaces, design schemes for battery swapping mechanisms, and installation positions of battery swapping systems on vehicles. Regarding software, four automotive industry standards have been developed to specify communication protocol requirements between vehicles, battery swapping systems, and battery swapping stations. This ensures a consistent language for interchange among all three entities.

3. **International Standard Research**

With the rapid globalization of Chinese automotive companies, the battery swapping model for EVs is also experiencing swift international development. Countries and regions such as Europe, the United States, and Japan have begun to promote and implement battery swapping systems, thereby establishing an industrial foundation for initiating and advancing international standards in battery swapping.

In the automotive sector, ISO/TC22/SC37—the subcommittee dedicated to EVs—has commenced research on international standards pertaining to battery swapping in EVs. China has proposed several international standardization research projects focused on battery swappable EVs. These include general requirements and application scenarios, interoperability protocols, and safety standards. Additionally, China has applied to establish relevant working groups aimed at conducting comprehensive international standardization research within the domain of battery swapping. In light of these efforts, extensive dialogues have been undertaken with major automobile-producing nations and organizations; progress regarding international standard projects and the establishment of working groups is currently underway.

In the field of infrastructure, IEC/TC69/WG13, the working group responsible for establishing safety requirements for EV battery swapping equipment, has undertaken research to develop international standards for EV battery swapping facilities. Three key standards have been published:

1. IEC TS 62840-1 "Electric Vehicle Battery Swapping System – Part 1: General Requirements and Guidelines"
2. IEC 62840-2 "Electric Vehicle Battery Swapping System – Part 2: Safety Requirements"

3. IEC PAS 62840-3 “Electric Vehicle Battery Swapping System – Part 3: Special Safety and Interoperability Requirements for Battery Exchange Systems Using Removable Battery Systems”

   These standards delineate design cases, safety requirements, and interoperability safety criteria between battery swapping systems and battery swapping stations. Currently, China is at the forefront of revising Parts 1 and 2 of IEC 62840.

4. **Research on Group and Local Standards**

   As the promotion of battery swapping technology advances, extensive research has been conducted on group standards. Currently, industry organizations such as the China Association of Automobile Manufacturers, the China Society of Automotive Engineers, and the China Electricity Council are actively engaged in researching and formulating group standards for battery swappable EVs and their associated infrastructure.

   These organizations have held comprehensive discussions regarding technical details, including the construction of shared battery swapping stations and various mechanisms for battery swapping solutions. This collaborative effort has resulted in the establishment of numerous group standards, such as “Construction Specifications for Shared Battery Swapping Stations for Electric Passenger Cars” and “Construction and Vehicle Technical Specifications for Shared Battery Swapping Stations for Electric Medium and Heavy Trucks.”

3.3.2 Construction of EV Battery Swapping Standard System

According to the technical framework of the EV battery swapping standard system, and in order to comprehensively address the standard requirements for the interchangeability among EVs, battery packs, and battery swapping stations, the architecture of this standard system is defined as comprising five key components: “general requirements,” “vehicle,” “battery swapping facilities,” “battery swapping interface,” and “service management.” A further subdivision of these five components—based on commonalities and differences in scope and technical specifications across each specific standard—yields eleven subclasses that are content-complete, structurally coherent, and clearly delineated (as illustrated in Figure 3.15, with numbers in parentheses representing system codes).

1. **Basic Common Standards**

   Basic common standards primarily govern the fundamental technologies within the EV battery swapping standard system.

   Terminology and definitions standards aim to unify essential concepts in the realm of EVs and infrastructure related to battery swapping. These standards provide consistent definitions for commonly used terms involved in the standard-setting process, ensuring standardized usage and consistency of meanings. This lays a foundation for a unified language across related industries and offers standardized terminology support for the development of other components of the standards.

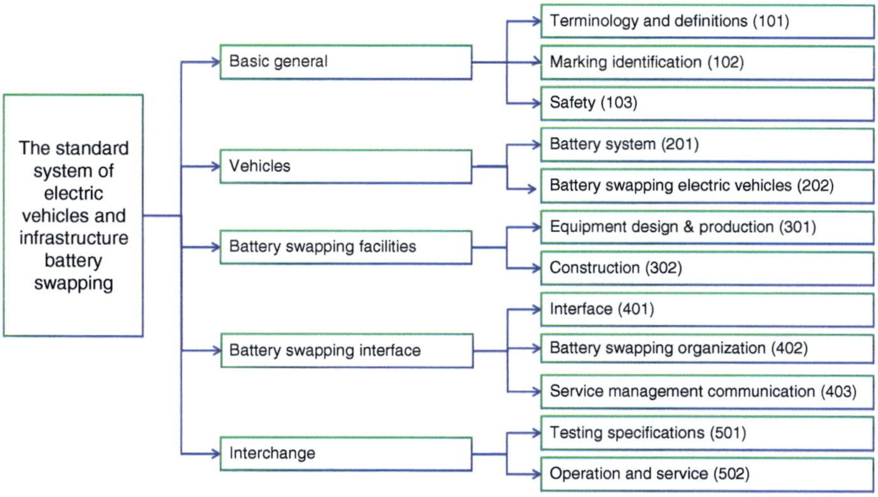

Figure 3.15 The standardization framework for battery swapping of electric vehicles and their infrastructure

Marking and labeling standards primarily delineate the standardization of terms, graphics, symbols, safety protocols, emergency response measures, etc., within the domain of EV and infrastructure battery swapping.

Safety standards encompass two primary components: EVs and battery swapping infrastructure. The standards applicable to the vehicle domain primarily address failure scenarios and establish safety requirements related to structural integrity, sealing protection, electrical faults, and environmental impacts during both the battery swapping process and the operational phases of battery swappable EVs.

In contrast, the standards pertaining to the infrastructure domain predominantly outline safety requirements concerning electrical safety and fire prevention during both the construction and operation of battery swapping stations.

2. **Vehicle Standards**

Vehicle standards primarily establish requirements concerning the interchangeability and various technologies associated with battery swappable EVs and their corresponding battery swapping systems.

Standards for battery swappable EVs mainly delineate the types of battery swapping technology routes (such as undercarriage swapping, rear-mounted swapping, side-mounted swapping, etc.), in addition to outlining general platform requirements, component layout schemes, and performance criteria for EVs across different technological pathways.

Standards pertaining to battery swapping systems predominantly specify the shape, structure, size requirements, quality ranges, installation positions of other components within the battery system, as well as any special performance criteria necessitated by the process of battery swapping.

3. **Battery Swapping Facility Standards**

   Technical standards pertaining to battery swapping stations primarily standardize activities such as planning and design, testing and inspection, construction, and operational services of these facilities.

   1. The planning and design standards for battery swapping stations encompass comprehensive technical management throughout the entire process, including feasibility studies, site investigations, design evaluations, engineering construction, and production operations. Additionally, they outline regulations regarding construction conditions as well as the selection and configuration of battery swapping stations. These standards propose relevant requirements concerning power station planning, site investigation methodologies, design specifications for power stations, energy conservation measures along with environmental protection assessments, technical supervision of specialized areas and key equipment within power stations, evaluation of operational indicators post-implementation, oversight in equipment manufacturing processes, project management protocols, and considerations related to both technical performance and economic viability.
   2. The testing and inspection standards delineate specific technical requirements alongside testing procedures for major equipment and systems utilized in battery swapping stations. Such standards include detailed criteria for essential components such as the structural integrity of various pieces of battery swapping station equipment, charging systems, control mechanisms governing the battery swapping process, monitoring systems, as well as stipulations for those battery swapping stations that are integrated into the electrical grid.
   3. Construction and acceptance standards articulate explicit requirements regarding the construction practices involved in building battery swapping stations. This includes guidelines on civil works along with installation procedures for system components within these facilities—covering aspects from initial start-up acceptance through to project handover—and culminates in a thorough evaluation of construction quality.

4. **Battery Swapping Interface**

   The battery swapping interface primarily encompasses the electrical interfaces, cooling interfaces, battery swapping mechanisms, and communication protocols essential for the coordination among battery swappable vehicles, battery swapping systems, and battery swapping stations.

   Electrical interface standards predominantly delineate the electrical performance, environmental adaptability, reliability, and other pertinent characteristics of the product. Additionally, they define aspects such as interface layout, shape, size, guiding circuits, mating parameters, and other relevant features of the electrical interface.

   Cooling interface standards mainly specify flow rate requirements along with environmental adaptability and reliability considerations. They also outline details regarding interface layout, shape, size specifications for flow requirements, mating parameters, and additional attributes related to the cooling interface.

Battery swapping mechanism standards primarily establish performance indicators, including clamping force and reliability of the battery swapping mechanism. Furthermore, they detail design types, critical dimensions, locking requirements, mating parameters, and other aspects of the battery swapping mechanism.

The standards for communication protocols are categorized into three distinct components: the interaction between the vehicle and the battery system, the exchange of information between the battery system and the battery swapping station, and the communication that occurs between the battery swapping station and the vehicle. These standards govern the language used for interactions among vehicles, battery swapping systems, and battery swapping stations. They delineate both signal senders and receivers while ensuring consistency and standardization in communication items and logic across all parties involved.

5. **Service Management**
   Service management primarily encompasses the requirements for interchangeability testing among vehicles, battery swapping systems, and battery swapping stations, as well as the operational and service standards applicable to battery swapping stations.

   Interchangeability test specification standards mainly govern the test equipment, testing procedures, and methodologies necessary to achieve interchangeability following the standardization of solutions between battery swapping systems, vehicles, and battery swapping stations. This includes specific protocols for conducting interchangeability tests between battery swapping systems and vehicles, as well as between battery swapping systems and their corresponding stations.

   Operational and service standards predominantly outline the requirements for network communication, data interaction, monitoring processes, electricity metering practices, power quality assessments, station-network interactions, along with other service activities and quality benchmarks pertinent to battery swapping stations.
6. **Research on the Development of Battery Swapping Standards**
   Going forward, standardization organizations for the automotive and power infrastructure sectors will enhance their cooperation and exchange of information. They will engage in comprehensive communication at three levels: the development of standard systems, coordination of standard projects, and dialogue on standard content. This collaborative effort aims to ensure the effective implementation of the battery swapping standard system, facilitate steady progress on relevant standard projects, and produce research outcomes that align with industry needs.

   Both parties will concentrate on establishing research standards related to safety requirements for battery swapping stations and battery swappable EVs. Additionally, they will focus on advancing chassis battery swapping technology for commercial vehicles, developing battery swapping systems, creating compatibility testing methods between vehicles and these systems, as well as facilitating information exchange regarding battery swapping services. This initiative seeks to further standardize and guide the evolution of the battery swapping industry while enhancing safety, interchangeability, and user convenience in battery swapping operations.

3.3.3 Challenges of Cross-regional Standard Interoperability

The development of battery swapping standards outside of China is characterized by fragmentation and progresses at a significantly slower pace compared to China's rapid advancements in this area. Various regions are investigating the potential benefits of battery swapping; however, they face numerous challenges, including high infrastructure costs, regulatory discrepancies, and a lack of standardization across different vehicle models.

3.3.3.1 Europe

Europe is gradually exploring the concept of battery swapping, primarily driven by Chinese companies such as NIO and energy firms like BP. NIO has already commenced the expansion of its swap station network into Europe, with plans to establish approximately 1,000 stations by 2025. However, adoption in Europe remains cautious, largely due to European automakers' skepticism regarding the economic viability of battery swapping

Standardization presents one of the most significant challenges; European manufacturers tend to prioritize advancements in battery range and charging speed over embracing a battery swapping model. Despite these obstacles, Norway and Germany have expressed interest in piloting battery swapping infrastructure. Norway, characterized by its high EV penetration, has witnessed early installations; however, the overall strategy across Europe continues to emphasize enhancing public charging infrastructure and fast-charging capabilities.

3.3.3.2 United States

In the United States, battery swapping has faced significant challenges in gaining traction. The country's extensive geography and dependence on private vehicles for long-distance travel render the logistics of battery swapping less appealing. Furthermore, American automakers such as Tesla have prioritized the development of a comprehensive charging network over pursuing battery swapping solutions. Although Tesla experimented with battery swapping in 2013, it ultimately discontinued this initiative due to insufficient customer demand.

Nevertheless, several startups and energy companies have recently rekindled interest in the concept, particularly for fleet and commercial vehicles where standardized designs and predictable routes could enhance the feasibility of battery swapping. For instance, Ample, a California-based company, is exploring modular battery swapping stations tailored for rideshare and delivery fleets. However, these efforts remain relatively small-scale when compared to China's expansive network.

3.3.3.3 India

India presents a more promising market for battery swapping, particularly in the context of two- and three-wheelers. The Indian government is actively promoting battery swapping as part of its broader initiative to advance electric mobility.

Battery swapping for electric rickshaws and scooters has demonstrated significant potential due to their lower battery requirements and standardized designs. However, similar to challenges faced in Europe and the U.S., large-scale adoption of passenger cars encounters obstacles related to the lack of standardization among vehicle manufacturers and high infrastructure costs.

3.3.3.4 Southeast Asia

Southeast Asia, similar to India, is actively exploring battery swapping solutions for two-wheelers (2Ws), which are predominant in the region's transportation landscape. Countries such as Indonesia and Thailand have expressed interest in this initiative, with local companies partnering with Chinese firms to establish battery swapping networks. The primary objective is to reduce emissions from urban mobility while providing affordable electric alternatives to gasoline-powered scooters and motorcycles.

While home charging infrastructure is well-established in many countries, the situation for 2Ws presents a distinct challenge. The stock and sales of 2Ws continue to rise in both India and ASEAN nations. These expanding markets are also witnessing increasing momentum in battery swapping technologies, particularly within India. In 2023, Gogoro, a Taiwan-based battery swapping company, announced a USD 2.5 billion partnership with the Indian state of Maharashtra. Gogoro plans to invest over USD 1.5 billion into deploying smart battery infrastructure that will include its network of battery swapping stations. Additionally, other Indian startups like Sun Mobility and Battery Smart have raised USD 50 million and USD 33 million, respectively, to further develop their battery swapping infrastructure.

In Africa as well, there has been a notable increase in investment in battery swapping technologies for 2Ws. Ampersand, a Rwanda-based company, currently facilitates approximately 140,000 monthly battery swaps for more than 1,700 customers who collectively travel around 1.4 million kilometers each week across Kigali and Nairobi. Furthermore, Spiro—a startup focused on electric 2Ws—secured around USD 60 million in funding during 2023 aimed at expanding its electric fleet and establishing over 1,000 additional swap stations.

The United States, India, and Southeast Asia each adopt distinct approaches to battery swapping, influenced by their specific transportation needs and market dynamics. While these regions may look to China's success in battery swapping as a potential model, their unique circumstances will dictate how they adapt this technology, resulting in significant differences in strategy and implementation.

3.3.3.5 United States: Small-scale, Fleet-oriented Efforts

In the United States, battery swapping has faced significant challenges in achieving widespread adoption, primarily due to the country's vast geography and consumers' preference for personal vehicles. Tesla's attempt to implement battery swapping in 2013 was unsuccessful, largely attributable to insufficient consumer demand. Since that time, attention has shifted toward fast-charging networks, particularly given Tesla's success with its Supercharger stations. Nevertheless, there is a resurgence of interest in battery swapping technology, especially within the fleet and commercial vehicle sectors, where it aligns more effectively with predictable routes and standardized vehicle designs.

Ample, a California-based startup, has developed modular battery swapping stations specifically designed for rideshare and delivery fleets. The company recognizes the potential of concentrating infrastructure around particular locations, such as densely populated urban areas or logistics hubs. In these settings, battery swapping could provide an effective solution to the lengthy recharge times that often render EVs less appealing for high-utilization applications like taxis and delivery trucks. However, the United States still lags significantly behind China's extensive network in this regard. Scaling up remains a challenge without increased government incentives or standardized regulations—efforts that China has actively pursued.

3.3.3.6 India: Battery Swapping for Two- and Three-wheelers

In contrast, India stands out as one of the most promising markets for battery swapping, particularly in the realm of two- and three-wheelers, which dominate the country's transportation landscape. The Indian government is actively promoting electric mobility through initiatives such as faster adoption and manufacturing of hybrid and electric vehicles (FAME) scheme, which endorses battery swapping as a viable solution for electric scooters and rickshaws. Unlike automobiles, these vehicles are equipped with smaller batteries, thereby facilitating easier and more cost-effective swapping processes.

Gogoro, a Taiwanese company, has taken advantage of this opportunity, announcing a USD 2.5 billion investment partnership with Maharashtra to establish smart battery infrastructure, including swapping stations. Additionally, Indian startups such as Sun Mobility and Battery Smart have secured significant funding—USD 50 million and USD 33 million, respectively—for expanding their battery swapping infrastructure. India's focus on electrifying urban mobility through shared, small-format EVs positions it to capitalize on battery swapping technology, much like China, although challenges related to standardization across vehicle manufacturers remain.

3.3.3.7 Southeast Asia: Targeting Urban Mobility

Southeast Asia, similar to India, is actively exploring battery swapping technology primarily for 2Ws. Countries such as Indonesia and Thailand are collaborating with Chinese companies to develop the necessary infrastructure for battery swapping, with a focus on reducing emissions from gasoline-powered scooters and enhancing urban air quality. For example, partnerships with firms like Gogoro are facilitating the deployment of battery swapping systems for electric scooters, providing an affordable and environmentally friendly alternative to traditional modes of transportation.

In Southeast Asia, motorcycles and scooters dominate the roads, making battery swapping a practical solution for these vehicles. By concentrating on smaller, low-cost vehicles that do not require large batteries, these countries are following China's lead in scaling up infrastructure for high-density urban mobility, where electric alternatives are most urgently needed.

3.3.3.8 Comparison to China's Strategy

In China, the government's top-down and aggressive strategy—characterized by the integration of national standardization for battery designs and extensive financial incentives—has facilitated a rapid expansion of battery swapping stations. Cities such as Beijing and Nanjing have been designated as pilot cities for this swapping model, with companies like Nio and BAIC reaping significant benefits from substantial government support. Furthermore, tax policies that exempt battery swapping vehicles from certain subsidy price caps have further accelerated their adoption.

In contrast, the United States continues to concentrate on commercial fleets while experiencing limited infrastructure development. Meanwhile, India and Southeast Asia have focused primarily on two- and three-wheeled vehicles—a sector that China has also supported but not prioritized to the same extent as four-wheeled vehicles. India's collaboration with international firms like Gogoro alongside local startups reflects China's model of public-private partnerships, albeit on a smaller scale. In all instances, standardization remains a significant challenge, particularly in markets where various vehicle types and manufacturers operate with differing battery systems.

Ultimately, the United States, India, and Southeast Asia are likely to pursue battery swapping initiatives tailored to their specific contexts: the United States will concentrate on fleet applications, India will advocate for the electrification of 2Ws, and Southeast Asia will capitalize on partnerships with international firms. While all regions may draw inspiration from China's achievements in this domain, their respective pathways to scaling battery swapping solutions will be shaped by their distinct transportation needs and market conditions.

Africa is emerging as a promising region for battery swapping, particularly in the 2W sector, following the examples set by India and Southeast Asia. Although Africa's electric mobility market remains in its early stages, battery swapping is gaining momentum due to the pressing need to reduce emissions and mitigate high fuel costs. Notably, countries such as Rwanda and Kenya are at the forefront of pioneering battery swapping initiatives for motorcycles and scooters, presenting a scalable solution for urban transportation that heavily relies on 2Ws.

In Rwanda, Ampersand stands out as a leader in battery swapping, operating one of the most extensive networks on the continent. As of 2023, Ampersand was conducting over 140,000 battery swaps monthly for more than 1,700 customers who collectively travel approximately 1.4 million kilometers each week across Kigali and Nairobi. This network enables electric motorcycle riders to exchange their depleted batteries for fully charged ones, thereby minimizing downtime and offering an economical alternative to gasoline-powered vehicles.

This network enables electric motorcycle riders to exchange their depleted batteries for fully charged ones, thereby minimizing downtime and offering a cost-effective alternative to gasoline-powered motorcycles. The success of Ampersand demonstrates that battery swapping is a viable solution in Africa, particularly when it addresses specific regional challenges such as high transportation costs and reliance on fuel.

In a similar vein, Spiro, another African startup, secured approximately USD 60 million in 2023 to expand its fleet of electric 2Ws and establish over 1,000 battery swap stations across the continent. Spiro's initiatives are instrumental in developing the necessary infrastructure to support EVs in urban areas where traditional EV charging facilities are scarce.

3.3.3.9 Comparison with China and Other Regions

Africa's approach to battery swapping mirrors that of India and Southeast Asia, with a primary focus on 2Ws due to the widespread use of motorcycles and scooters as essential modes of transportation. Similar to India, where battery swapping for electric rickshaws and scooters has gained traction owing to smaller, more standardized batteries, Africa is leveraging the compact size and lower costs associated with 2Ws to expedite its transition toward electric mobility.

In contrast to China, where battery swapping is being implemented across various vehicle categories—including passenger cars and heavy-duty trucks—Africa's strategy remains more specialized, addressing its unique urban mobility challenges. Chinese companies such as Nio have benefited from top-down government standardization coupled with robust financial incentives; in contrast, African nations are increasingly relying on private startups and international collaborations to establish their infrastructure. Furthermore, Africa's battery swapping initiatives are evolving without the extensive government intervention observed in China; however, there is a growing interest in expanding this sector through foreign investment.

3.3.3.10 Challenges and Opportunities

One of the primary challenges facing Africa is the insufficient development of widespread infrastructure and standardization. In contrast to China, where the government has implemented national battery standards to facilitate battery swapping, Africa's market remains fragmented, with various startups offering competing systems. Nevertheless, there exists significant potential for growth, as battery swapping presents a more cost-effective and sustainable solution compared to establishing an extensive charging network—particularly in countries where access to electricity is still limited.

Moreover, battery swapping could effectively address issues related to the reliability of electricity grids in Africa. By creating a network of swapping stations, companies can operate independently from unreliable grid infrastructures; fully charged batteries can be stored and exchanged without requiring a continuous power supply—a critical advantage in many African cities.

In conclusion, Africa is charting a course akin to that of India and Southeast Asia, where battery swapping is emerging as a pivotal solution for the electrification of urban transport, particularly concerning 2Ws. Companies such as Ampersand and Spiro are at the forefront of this movement, possessing the potential to revolutionize urban mobility across the continent by offering affordable and environmentally friendly alternatives to traditional fossil fuel-powered vehicles. While Africa continues to grapple with challenges related to infrastructure and standardization, battery swapping presents a significant opportunity for accelerated EV adoption, especially within its cities.

4

Business Models and Case Studies of Battery Swapping: A Comprehensive Analysis

4.1 Two-wheeled and Three-wheeled Electric Vehicles: High-frequency and Essential Use Cases

4.1.1 Business Models: Subscription vs. Pay-per-use

Worldwide, the rising environmental awareness and increasing demand for sustainable transportation solutions have led to significant attention on electric two-wheelers (E2Ws) and three-wheelers (E3Ws) as clean, efficient, and economical modes of transport. Government policies—such as subsidies and tax reductions—coupled with technological innovations like advancements in battery technology and improvements in charging infrastructure, have further accelerated the rapid development of the electric two-wheeler and three-wheeler markets. In recent years, the global two-wheeler market has expanded notably, particularly in Asian regions such as China, India, and Southeast Asia, establishing itself as a primary mode of daily transportation.

Global sales of E2Ws reached 59 million units in 2022, with projections indicating growth to 98 million units by 2026. The global electric bicycle market was valued at $37.47 billion in 2022 and is expected to reach $119.72 billion by 2030, reflecting a compound annual growth rate (CAGR) of approximately 15.6%. The lithium-ion battery segment continues to dominate this market.

As for the global E3W market size, it stood at $2.71 billion in 2023 and is forecasted to grow to $12.81 billion by 2032, exhibiting a CAGR of about 18.0%. This market primarily comprises passenger and cargo segments; vehicles capable of traveling up to 100 km are predominant within this sector.

While the charging infrastructure is improving, there remains a significant shortage of charging stations in certain regions, particularly in developing countries. This limitation hinders the adoption of electric two- and three-wheelers. The initial cost of these vehicles is generally higher than that of traditional fuel-powered alternatives, especially when battery prices are elevated, creating a barrier for middle- and low-income consumers. Furthermore, safety standards and usage regulations vary across different countries and regions, posing challenges for multinational corporations.

Advancements in battery technology are expected to significantly enhance the range of electric two- and three-wheelers while reducing charging times. The integration of smart features will further elevate the user experience. As production scales up and technology matures, it is anticipated that the costs associated with electric two- and three-wheelers will decrease, thereby increasing accessibility for consumers. As the market evolves, governments worldwide are likely to continue implementing supportive policies aimed at unifying international standards to foster healthy market development.

In addition to traditional short-distance commuting and cargo transport applications, E3Ws are poised to find broader uses within last-mile delivery services and sharing economy sectors. Utilizing more eco-friendly materials along with sustainable manufacturing processes—coupled with improvements in energy efficiency—are long-term objectives for the industry.

China's E3W industry has undergone three distinct phases of development since 2001: the initial development phase (2001–2005), the rapid growth phase (post-2006), and the adjustment period following 2017, during which environmental policies have intensified competition within the sector. Globally, both E2W and three-wheeler markets are exhibiting similar trends, driven by technological advancements and increasing environmental awareness that position these vehicles as essential components of urban transportation. Figure 4.1 shows the sales volume of electric two-wheeled electric vehicles in China (2015–2024).

In China, the market for E3Ws expanded from 25.724 billion yuan in 2017 to 39.827 billion yuan in 2022, reflecting a consistent rise in demand and production levels. In India, registrations of E2Ws surged by 24% year-on-year in February 2024, achieving a market penetration rate of 5.7%. Projections indicate that by 2030, approximately 80% of two-wheelers will be electrified. Furthermore, Thailand's electric motorcycle market demonstrated robust growth in 2023, with sales increasing by an impressive 117.2%.

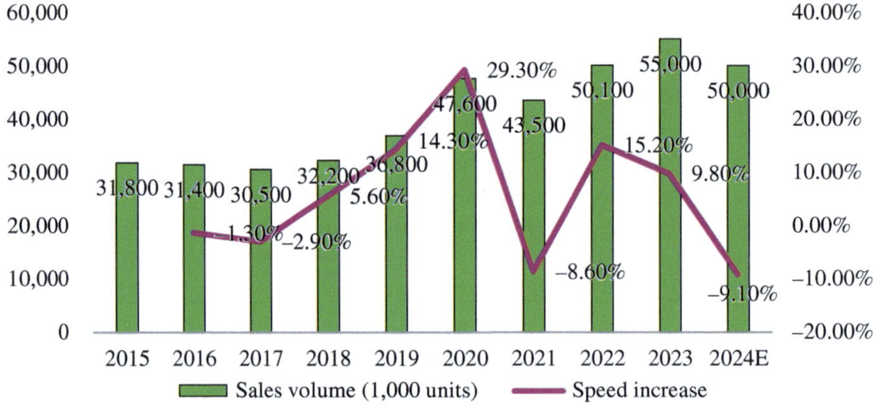

Figure 4.1 Sales of electric two-wheelers in China (2015–2024). *Source:* Giliomee, J; Booysen, M. J, Decarbonising South Africa's long-distance paratransit: battery swapping with solar-charged minibus trailers. *Transportation Research Part D: Transport and Environment* 2023, 117, 103647.

Currently, in the industry, rental and battery swapping platforms generally adopt two primary models: direct operation and the deployment of battery swapping cabinets. A limited number also utilize a franchise model that combines service outlets with the installation of battery swapping cabinets.

Direct operation model: In this model, enterprises independently establish battery swapping stations, procure batteries, and offer battery swapping services to users. The advantage of this approach lies in centralized management and maintenance of both the stations and batteries, which enhances service quality and operational efficiency. However, a notable disadvantage is that both construction and operational costs tend to be relatively high.

Cooperation model: This model involves collaboration between enterprises or institutions to jointly develop battery swapping stations while providing services for users. The key benefit here is the potential reduction in construction and operational expenses alongside improved resource utilization. Conversely, a significant drawback is that effective coordination of interests between cooperating parties is essential to prevent conflicts or disputes.

Franchise model: Under this framework, enterprises grant authorization for their battery swapping business to franchisees who then construct their own stations and deliver services to users. The primary advantage of this model is its ability to rapidly scale operations and enhance market share. Nonetheless, an inherent challenge lies in the variability of management practices among franchisees; inconsistencies may adversely impact the brand image of the parent enterprise. The overall industrial value chain, spanning from upstream raw materials to downstream end-users, is illustrated in Figure 4.2.

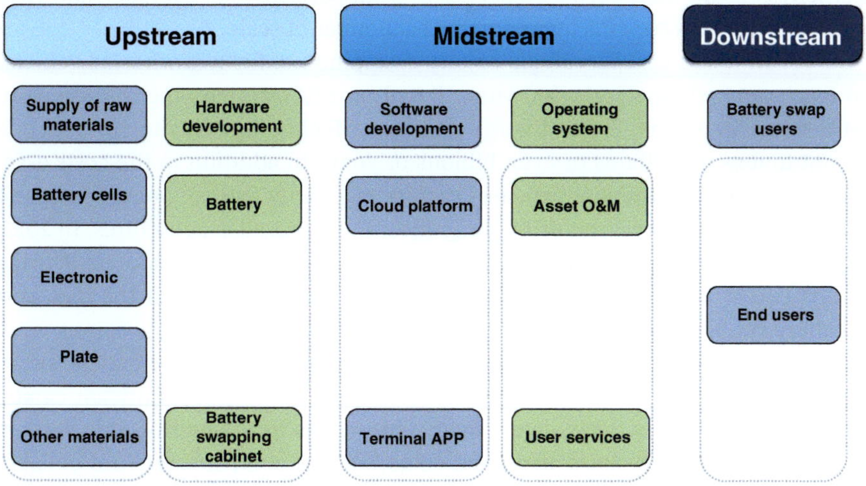

Figure 4.2 Schematic diagram of the industrial chain of the electric two-wheeled vehicle battery swapping industry.

4.1.1.1 Profit Model of Two-wheeled Electric Vehicle Battery Swapping Project

In terms of investment, the capital allocated to two-wheeled electric vehicle battery swapping cabinets primarily encompasses two key aspects: products and operations. The product composition includes batteries, cabinets, and systems, while the operational components consist of labor costs, electricity fees, site rental charges, and other related expenses.

Secondly, regarding profitability, the principal source of income from two-wheeled electric vehicle battery swapping is derived from rental fees paid by riders. It is important to note that variations in cities, demographics, timeframes, locations, and available resources lead to differing investment outcomes and returns.

For instance, in second- and third-tier cities across China, the monthly rent for a two-wheeled electric vehicle battery swapping cabinet averages around 300 yuan. After accounting for depreciation costs associated with batteries and cabinets—resulting in an approximate profit margin of 200 yuan—the net profit per customer order typically ranges between 100 and 150 yuan when further operational expenses are deducted. Overall, this results in a comprehensive profit margin estimated at approximately 40–50%.

Thus far we can establish a foundational financial model concerning our investments versus outputs. A critical factor here is that user numbers directly influence our output levels while available resources dictate overall costs. How can we effectively manage these costs? What lies at the core of battery swapping? The essence of battery swapping revolves around leasing batteries followed by facilitating unlimited exchanges. Maximizing both the battery rental rate and asset utilization rate emerges as a pivotal element influencing the payback period for our two-wheeled electric vehicle battery swapping initiative.

From a business perspective, the implementation of a battery swapping model facilitates the intensive utilization of batteries and enables effective management of electricity consumption during peak and off-peak periods. This approach can yield significant business benefits by leveraging economies of scale.

The construction cost of a single battery swapping site for two-wheeled vehicles can reach approximately 110,000 yuan. Based on this estimate, the initial investments made by China Tower Energy and Yugu Technology in site construction amount to tens of millions or even billions of yuan. In addition to the costs associated with site construction, expenses such as electricity fees, site rental fees, and customer acquisition costs contribute significantly to overall expenditures. In 2022, Yugu Technology incurred over 100 million yuan in electricity fees alone. During the first half of 2023, electricity expenses still represented 38.67% of the total cost for battery swapping services.

The hardware requirements inherent in the two-wheeled vehicle battery swapping model result in substantial asset-heavy characteristics and significant capital investment, which effectively deters many enterprises from entering this market. To mitigate these heavy asset attributes, some internet companies opt for collaborative models with hardware manufacturers while primarily focusing on software development. This partnership approach can lower entry barriers to some extent and alleviate financial pressures; however, it also introduces new challenges.

This cooperative model tends to elongate the supply chain: on one hand, it complicates fund management; on the other hand, poor collaboration among multiple enterprises and agents may lead to issues such as site closures, subpar battery quality, and delayed after-sales service—factors that could tarnish reputation within rider communities and subsequently impact business performance negatively.

Even if stable operations and services are maintained over time, inconsistent revenue streams will continue to exert pressure on businesses within this sector. Notably, even larger-scale enterprises operating in the two-wheeled vehicle battery swapping industry do not achieve substantial revenue or profit margins.

The underlying process of the two-wheeled vehicle battery swapping business can be primarily divided into four components: "standardized lithium battery modules + battery swapping cabinets + mobile phone applications + remote data platforms." Each component necessitates the support of ecological thinking. Currently, the entire two-wheeled vehicle battery industry remains in a phase characterized by independent research and development conducted by major manufacturers. Consequently, the degree of standardization in battery swapping is notably low.

On one hand, this lack of standardization is evident in the varying battery standards among different manufacturers; these standards differ significantly regarding specifications such as materials and energy density. On the other hand, it manifests in non-standardized battery compartments across various electric vehicle brands. Collectively, these two factors contribute to incompatibility between batteries intended for swapping and vehicles themselves, leading to a direct issue—escalated user costs.

Presently, evaluations from takeout riders concerning two-wheeled vehicle battery swapping reveal a polarized trend. One group offers highly favorable assessments of this service, asserting that it alleviates their anxiety and reduces costs. Conversely, another group perceives the expenses associated with two-wheeled vehicle battery swapping as excessively high, likening it to an impractical option.

The cost implications of battery swapping are largely attributed to its non-standardized nature. Riders often find that their own electric vehicles do not align with nearby battery swapping cabinets; thus they face a dilemma—either invest in a new compatible electric vehicle or purchase an additional spare battery outright. The price range for an E2W typically falls between 2,000 and 3,000 yuan, while a lithium battery costs approximately 1,300 yuan—making it more economical than acquiring an entirely new vehicle.

Therefore, many riders may find themselves torn between the two options and might even abandon the "battery swapping" model altogether. For manufacturers of battery swapping systems for two-wheeled vehicles, losing users due to a lack of standardization is a regrettable setback that hampers both their growth and that of the entire industry.

In terms of policy, the government has been actively promoting the standardization of battery swapping technologies. The "Technical Specifications for Centralized Charging Facilities for Electric Bicycles - Part 1," which was implemented last year, specifies that battery swapping cabinets are only applicable to charging and exchanging 48 V lithium-ion battery packs with rated voltage. Additionally, it outlines technical requirements for these cabinets, encompassing 36 subdivision parameters across 14 categories.

The market landscape and regulatory frameworks have established clear demands for standardization among manufacturers engaged in two-wheeled vehicle battery swapping. If various manufacturers adopt an ecological mindset and further advance efforts toward achieving standardization, it will not only benefit their own development but also foster progress within the industry as a whole. Conversely, those manufacturers unable or unwilling to adapt to standardized practices are likely to face obsolescence.

Ecological thinking should not only be integrated into the battery standardization process but also in the "mobile phone APP + remote data platform" aspect.

Once hardware facilities, such as battery swapping cabinet sites, are established, the software experience becomes crucial in determining the competitiveness of battery swapping manufacturers. For instance, can an optimal battery swapping plan be selected based on a rider's delivery route?

These considerations must be addressed within the "mobile phone APP + remote data platform" framework, which represents a significant advantage for Internet enterprises—particularly those that focus on local life service platforms. In 2019, China Tower Energy and Meituan Delivery initiated an in-depth collaboration to provide battery swapping services for electric vehicles used in food delivery.

This partnership has the potential to create a win-win scenario: China Tower Energy can assist Meituan riders by addressing their challenges related to battery charging and swapping. Meanwhile, Meituan Delivery can offer substantial order volume and robust data capabilities to China Tower Energy, thereby enhancing user experience at the software level for riders.

Currently, alongside launching its own "Yuanyuan Battery Swapping," Meituan is also collaborating with other two-wheeled vehicle battery swapping companies. Beyond capital investment in start-ups, there exists a group of Internet enterprises that may emerge as players entering the two-wheeled vehicle battery swapping market. Herein lies an intriguing speculation: these could include local service providers such as Amap, DiDi, and Energy Monster.

At present, DiDi has already ventured into the realms of takeout and errand services. The company may further expand its offerings by entering the two-wheeled vehicle battery swapping market.

Monster Charging, a prominent player in the shared power bank sector, is experiencing a continuous decline in performance. In the third quarter of 2023, its revenue amounted to 614 million yuan, reflecting a decrease of 24.7% compared to 815 million yuan during the same period in 2022. Compounding this concern is the perception among investors that two-wheeled vehicle battery swapping is poised to "supplant" shared power banks as the next focal point within the sharing economy.

Looking ahead, it is inevitable that two-wheeled vehicle battery swapping services will extend beyond just takeout riders and gradually permeate into broader public usage. In this evolution, only through win–win collaborations can we achieve wider user coverage and enhance user experience—this aligns with the long-term business strategy anticipated by battery swapping manufacturers.

From an international perspective on battery swapping scenarios, ADL's analysis indicates that B2B segments for E2Ws (electric two-wheelers) and E3Ws (electric three-wheelers) currently offer significant opportunities. However, low profitability remains a pressing issue since revenue primarily stems from battery-swapping fees while incurring substantial expenditures related to high capital expenditure (CAPEX), spare battery inventory management, as well as operational and maintenance costs. To establish a viable business model, ADL recommends that operators adopt a dual approach focused on maximizing revenue while minimizing costs, as detailed below.

1. **Maximize Revenue**
   - Secure captive demand: Partner with fleet operators or operate a fleet business to ensure captive demand, resulting in assured off-take.
   - Flexible pricing schemes: Adjust fee structures based on customer type and preference, either on a per swap, per kilometer, or subscription model basis to enable top-line maximization.
   - Multiply revenue streams: Utilize multi-channel income sources such as advertising at swapping centers and enter the battery recycling business over the medium term. Generate income through the extraction of spent materials in the batteries. Position swapping stations close to convenience stores to increase foot traffic and bring additional revenue to the store. Charge a commission for increased sales.
2. **Minimize cost**
   - Partnership and hybrid models: Implement a joint partnership or hybrid model to reduce investment in batteries and swapping infrastructure. Explore the feasibility of a franchise model and configure the business portfolio toward an asset-light model by considering leasing the swapping technology and exploring partnerships with local stores.
   - Interoperability: Operating in B2B segments, especially the fleet business, would allow minimizing the different types of battery inventory and keeping it at an optimum level.
   - Focused expansion of station: Launch stations in high-density areas to ensure sufficient take-off and profitability. Consider expansion only if there is proven demand for off-take in the near future.
   - Use incentives for electricity and physical infrastructure: Lobby with governments for a subsidized cost of electricity, which is close to 35–40% of the cost of a swapping station. Similarly, governments may offer physical infrastructure at a subsidized cost, which is typically 20–22% of the cost of a station.

Potential market entrants are encouraged to assess the feasibility of battery swapping within the B2B segments of E2Ws and E3Ws. When contemplating entry into the battery swapping sector, it is essential to consider the aforementioned levers for "maximizing revenue/minimizing cost." A viable market opportunity exists, provided that a well-defined market-entry strategy and target-segment approach are implemented.

Currently, the further development of the two-wheeler electric vehicle battery swap industry is hindered by several factors, including inadequate charging infrastructure, inconsistent battery swap standards, and underdeveloped battery technology. Different brands of electric vehicles feature varying battery swap interfaces, which prevents users from exchanging batteries at stations operated by different manufacturers. Furthermore, there are ongoing concerns among users regarding the endurance and lifespan of electric vehicle batteries, highlighting a need for improvement in these areas.

To facilitate the unification of industry standards, government entities can enhance investment in the construction of charging facilities and promote the establishment of additional battery swap stations. Additionally, it is essential to formulate relevant standards and regulations that will encourage orderly growth within the industry. Electric vehicle brands should strengthen collaboration and engage in negotiations to jointly establish standardized battery swap interface protocols. Moreover, advancements and innovations in battery technology are crucial for standardizing practices across the sector. Companies can increase their investment in research and development efforts aimed at enhancing both battery endurance and lifespan to better meet user demands.

4.1.1.2 High Initial Capital Investment for Two-wheeler Swap Stations

The establishment of battery swap stations for two-wheelers entails not only expenses associated with land acquisition, labor, power grid modifications, swap cabinets, and other necessary equipment but also significant costs related to battery production and electricity consumption. Industry estimates suggest that the investment required to construct a single two-wheeler battery swap station is approximately 110,000 yuan.

Publicly available data indicates that China Tower, currently the largest operator in terms of user base, manages around 50,000 battery swap stations. Costrip has experienced rapid expansion and has deployed nearly 20,000 stations within just one year. Additionally, Hellobike Battery Swap and Zhiru Battery Swap have each established over a thousand stations in major cities such as Shanghai and Changsha. The initial capital investments made by these brands for setting up their respective swap stations reach tens of millions or even billions of yuan.

Furthermore, the heavy asset nature of the two-wheeler battery swap model extends beyond hardware to include the development of supporting software. This aspect necessitates significant capital investment and represents a relatively weak link in the current industry's capital management. Most existing battery swap brands operate as internet companies responsible for software development while collaborating with hardware manufacturers. Such a multi-party cooperation model inevitably lengthens the supply chain, complicating effective capital management and potentially resulting in challenges such as the closure of swap stations, delays in battery swapping, and server downtime due to inefficient collaboration among various companies and agents.

This indicates that enterprises entering the battery swap market should prioritize shortening their middle supply chain. Models like Costrip's full-industry-chain self-operation, Yadea's vehicle-battery integration, and Ant Financial's investment plus operation model with Hellobike can enhance capital control and streamline the supply chain, thereby fostering an environment conducive to healthy brand development.

4.1.2 China Tower Corporation (Rural and Urban Coverage)

China's new energy vehicle (NEV) industry has evolved from an emerging stage to a mature one, establishing a leading position on the global stage. According to statistics from the China Association of Automobile Manufacturers (CAAM), production and sales of NEVs (including commercial vehicles) in 2023 reached 9.587 million and 9.495 million units, respectively, reflecting year-on-year growth rates of 35.8% and 37.9%. The penetration rate of NEVs climbed to 31.6%, representing an increase of 5.9 percentage points compared to the previous year (see Figure 4.3).

For nine consecutive years, China has maintained its status as the world's largest NEV market, accounting for over 60% of global NEV sales in 2023. These accomplishments signify a new phase for Chinese NEVs as they enter the mass market; industry standards are gradually being unified while the market continues to mature.

By the end of 2023, there were approximately 20.41 million NEVs on China's roads, constituting about 6% of the total vehicle population; among these, around 15.52 million were pure electric vehicles, which represent approximately 76% of the overall NEV stock.

China Tower Energy Co., Ltd. was established in June 2019 as a wholly owned subsidiary of China Tower Corporation. In the same year, China Tower officially launched its battery swapping network through its subsidiary, China Tower Energy Co., Ltd., providing low-speed electric vehicle battery swapping services specifically tailored for couriers in the food delivery and logistics sectors.

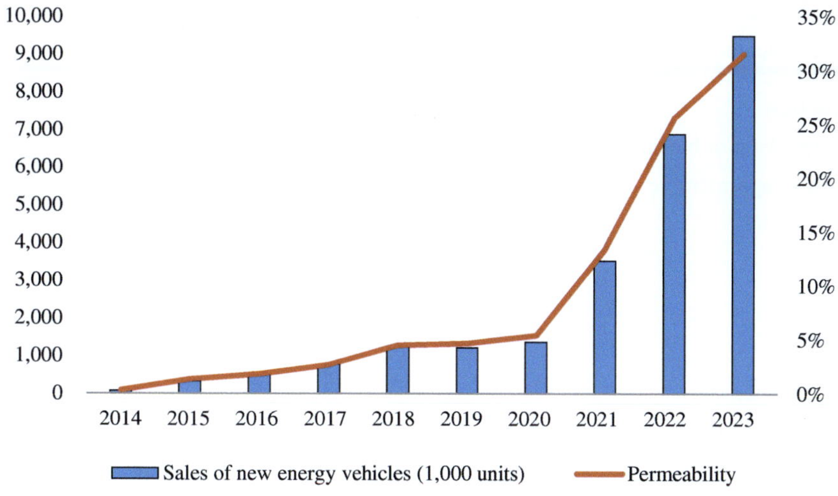

Figure 4.3 China's New Energy vehicle sales (in thousand units) from 2014 to 2023.

China Tower Battery Swap utilizes standard lithium-ion batteries as carriers and integrates advanced technologies such as mobile IoT, cloud platforms, deep inspection algorithms, safe and intelligent charging solutions, GPS, among others. This enables the creation of a comprehensive intelligent management platform that encompasses low-speed electric vehicles, batteries, and operators. The service primarily caters to couriers operating within the logistics and food delivery industries.

In terms of its battery swapping business model, China Tower Energy is committed to accelerating product iteration and upgrades while enhancing its service system infrastructure. The company continues to explore opportunities within the consumer market (C-end) while rapidly expanding its clientele among corporate customers (B-end). As a result, the user base has experienced significant growth; by December 31, 2023, China Tower Battery Swap had reached 1.145 million users—an increase of 243,000 from the previous year—further solidifying its leading position in the light electric vehicle battery swapping market. The key technical features and advantages of such smart systems, in comparison to traditional alternatives, are detailed in Table 4.1.

4.1.2.1 Battery Swap Types

Shared battery swap mode: China Tower provides batteries, battery swap cabinets, and a swap platform in public areas, offering unlimited usage on a monthly subscription basis.

Service targets users who frequently use low-speed electric vehicles, such as food delivery riders.

Exclusive rental mode: China Tower offers core lithium batteries, swap cabinets, swap platforms, integrated charging and swapping stations, and other equipment for exclusive use by customers.

Table 4.1 The detailed performances of smart battery swapping.

Mode comparison	Smart battery swapping
Function	Lithium battery
Comparison of battery life	Stable battery life
Normal operating temperature range	−10 °C to +55 °C
Battery life	Permanent preservation
Security	Short-circuit protection, over-current protection, no-load protection, overload protection, full self-stop
BMS	Yes
Location features	Yes
Remote monitoring	Yes
Trajectory	Yes

Service targets users who frequently utilize low-speed electric vehicles in industries such as express delivery and logistics.

1. **Shared Battery Swap Mode:**
   Before the swap: Food delivery riders relied on lead-acid batteries that were expensive to purchase, experienced annual performance degradation, required lengthy charging times—which posed safety risks—and ultimately failed to meet daily delivery efficiency demands.
   After the swap: With the convenient battery swap service provided by China Tower Energy, food delivery riders save considerable charging time, enabling them to complete an additional 5–6 orders per day. At an estimated earnings of approximately 6–7 yuan per order, this results in a monthly income increase of at least 1,000 yuan for the riders while also addressing issues such as battery theft and unsafe charging practices.
2. **Exclusive Rental Mode:**
   Before leasing: Customers utilized lead-acid batteries that were not only costly but also exhibited poor performance, exceeded weight limits, and had limited ranges.
   After the lease: China Tower Energy offers battery leasing services that not only reduce the costs associated with battery purchases but also enable customers to take advantage of electronic fencing and GPS monitoring functionalities provided by its battery services. This capability facilitates real-time location tracking, speed monitoring, vehicle recovery in cases of theft, and analysis of vehicle behavior, thereby delivering valuable data support for optimizing delivery efficiency.

4.1.3 Zheli Power Swap (Instant Distribution Service)

Yugu Technology is a prominent domestic provider of charging and battery swap equipment and services specifically designed for E2Ws. The company aspires to establish itself as a world-class smart energy technology enterprise, guided by the business philosophy of "enhancing the safety, convenience, and intelligence of battery swaps for riders." Yugu Technology focuses on improving safety, efficiency, and user experience while actively responding to national policies regarding safe charging for E2Ws. The company innovates in smart energy technology, reshapes new energy consumption models, and leverages its independently developed smart hardware along with end-cloud collaboration, big data analytics, artificial intelligence (AI), and other core technologies to deliver safe, intelligent, and convenient charging and battery swap products and services.

The primary offerings of Yugu Technology encompass battery swap services as well as charging and battery swap equipment.

Battery swap services involve the provision of comprehensive solutions through self-developed software and hardware that includes lithium batteries, battery swap cabinets, and swapping systems. End users can operate these services via mobile applications or other client interfaces to facilitate offline self-service battery swapping. Under this service model, terminal users are not required to purchase batteries outright. Upon their initial use of the service, users must register an account for identity verification purposes before purchasing a package; they then scan a QR code to retrieve a fully charged battery from the designated swap cabinet. For subsequent uses, users can locate nearby swap cabinets using the mobile app; they simply scan a code to return their used battery before retrieving another charged one from the cabinet in order to continue operating their vehicle. Typically, each individual battery swap takes less than one minute.

Charging and battery swap equipment encompasses both swapping apparatus (including swap cabinets) as well as charging infrastructure (such as charging stations).

"Zheli Battery Swap" is the battery swap service application developed by Hangzhou Yugu Technology Co., Ltd. As a dedicated manufacturer and service provider of intelligent charging and battery swapping equipment for E2Ws, Yugu Technology has made significant strides since entering the electric bicycle battery swap market in 2019. The company's revenue has surged from over 87 million yuan to 220 million yuan, and further to 550 million yuan by 2023, while net profits have increased dramatically from over 15 million yuan to more than 81 million yuan.

As a pioneer in the battery swap model for E2Ws, Yugu Technology has amassed extensive experience in this field. Leveraging its technological advancements in smart hardware, end-cloud collaboration, big data analytics, and artificial intelligence (AI), the company has integrated an advanced data information system into its battery swap equipment. This system is capable of sensing, transmitting, receiving, storing, analyzing data, and making informed decisions based on that data—effectively providing the swapping equipment with a "nervous system" and "brain." These innovations enhance safety and efficiency while significantly improving user experience; AI replaces human expertise through predictive maintenance rather than reactive supervision and automates control processes instead of relying on manual labor.

By employing dynamic collection methods alongside efficient transmission techniques and precise automated decision-making across multidimensional spatial datasets as well as multi-time series data streams, Yugu Technology delivers increasingly digitalized, informationally rich, automated solutions for battery swapping services. This positions them with a formidable competitive advantage within the industry. By the end of 2022, Yugu's battery swap operations had been established across 58 cities—including Beijing, Shanghai, Guangzhou, Shenzhen, Hangzhou—and Chengdu—with a total deployment of approximately 20,500 battery swap cabinets housing around 400 thousand batteries online to serve approximately 291 thousand users.

In terms of service efficiency, Yugu Technology is committed to innovating in big data, the Internet of Things (IoT), and end-cloud collaboration technologies. These advancements are integrated throughout the battery swap service process to optimize resource allocation and enhance synergies, ultimately improving the overall efficiency of battery swap services.

It has been reported that Yugu Technology has developed "key technologies for large-scale energy charging and battery swap network optimization based on end-cloud collaboration." This development is founded on three core technologies: low-latency and high-reliability end-cloud collaboration; a general-purpose intelligent lithium battery model deployed on the device side; and intelligent resource allocation and scheduling for charging and battery swapping.

For Yugu Technology, harnessing artificial intelligence to address challenges such as efficient collaboration in end-cloud environments and the allocation and scheduling of resources within charging and battery swap networks significantly enhances service efficiency. This improvement not only boosts customer satisfaction but also strengthens market competitiveness. Additionally, it opens avenues for innovative business models and value-added services, thereby establishing a foundation for sustainable development.

In terms of quality service, Yugu Technology prioritizes product innovation driven by actual user needs. The company is actively engaged in the development and application of large-scale lithium battery models to deliver high-quality battery swap products and services.

Within the realm of battery swap services, the quality of the battery itself is paramount, serving as a reflection of the brand's commitment to excellence. Recently, Yugu Technology introduced the industry's first general-purpose intelligent lithium battery large model on device platforms. According to relevant authoritative institutions, various anomaly detection models based on this lithium battery large model achieve an accuracy rate exceeding 98%, with a capacity estimation error margin of less than 3%. This advancement effectively addresses core technical challenges related to lithium battery anomaly detection and range prediction.

The two-wheeled electric vehicle battery swap market has entered a competitive phase. It is essential to meet user needs and achieve sustainable development by offering high-quality products and efficient services. Zheli Battery Swap fosters growth through quality and efficiency, while simultaneously advancing the development and progress of the electric vehicle battery swap industry.

The E2W battery swapping industry is dominated by a few leading battery swapping operators, primarily including China Tower Energy, Yugu Technology, eSwap, and Zhizu Battery Swap. These companies have accumulated a certain user base and industry experience in the battery swapping sector. A comparative overview of the key operational data, market position, and technical strength of these leading companies is provided in Table 4.2. In addition to these major players, there are also several regional mid-sized and small battery swapping operators in the market.

4.1.4 Risks and Countermeasures for Enterprises in the Standardization Process of Two-wheeler Battery Swaps

4.1.4.1 Risks Faced by Enterprises in the Process of Standardization of Battery Swaps

At present, the standardization level of battery swapping for two-wheelers is relatively low. This is evident in the varying specifications of battery standards, including materials and energy density, among different brands. Additionally, there is inconsistency in the sizes of battery compartments across vehicles produced by

Table 4.2 Comparison of key data and technical strength of battery swap companies.

Company name	Key business data, indicators	Market position	Technical strength
Tieta Energy Co., Ltd.	By the end of 2022, operated 57,000 battery swap cabinets, with 902,000 online users, achieved revenue of 1.8 billion yuan from battery swap services in 2022	Industry's first-tier	According to the inquiry of the China and multinational patent information website, as of December 31, 2024, it had 575 invention patents
Shenzhen Yima Technology Co., Ltd.	By the end of 2020, had deployed over 8,000 battery swap cabinets	Industry's first-tier	According to the inquiry of the China and multinational patent information website, as of December 31, 2024, it had 195 invention patents
Shanghai Zhizu IoT Technology Co., Ltd.	Not disclosed	Industry's first-tier	According to the inquiry of the China and multinational patent information website, as of December 31, 2024, it had 168 invention patents
Yugu Technology	By the end of 2022, the company operated 2,050 battery swap cabinets, with 291,700 online users	Industry's first-tier	According to the inquiry of the China and multinational patent information website, as of December 31, 2024, it had 97 invention patents

various manufacturers. These two factors contribute to challenges regarding the compatibility of battery swap batteries with vehicles. Consequently, policy guidance is essential to further develop and implement standardized practices for battery swaps, transitioning from industry norms to nationally mandated standards.

On December 30, 2022, the State Administration for Market Regulation and the National Standardization Management Committee jointly issued a document titled "Technical Specifications for Concentrated Charging Facilities for Electric Bicycles - Part 1." The release of this regulatory document provides robust support for advancing standardization within the two-wheeler battery swap industry. Furthermore, these forthcoming regulations will impose more stringent production and technical requirements on enterprises involved in this sector. Currently, the entire E2W and battery swap industry remains in a phase characterized by independent research and development conducted by major manufacturers. The establishment of unified national standards will necessitate that enterprises fully standardize their equipment, processes, and production methods; noncompliant battery swap equipment may face obsolescence as a result.

The implementation of the "Technical Specifications" may have a significant impact on brands with noncompliant products. However, leading companies such as China Tower, Costrip, and Aihuanhuan each possess distinct advantages that mitigate these effects. For instance, China Tower, being a state-owned enterprise, is adept at seizing policy opportunities and benefits from certain inherent advantages. The Yadea & Aihuanhuan brand excels in integrating whole vehicles with battery swap technology, thereby achieving production efficiencies through vehicle standardization. Additionally, Costrip's full-industry-chain self-operation model enables it to independently manage the development and deployment of compliant equipment in accordance with regulatory requirements. This strict adherence to self-production and operational guidelines affords it notable production capabilities and profit margins. By leveraging their respective strengths to navigate risks effectively, these brands are well-positioned to establish robust defenses during the ongoing standardization process within the two-wheeler battery swap industry.

In light of the rapid advancements occurring in the two-wheeler battery swap sector, enterprises must prioritize capital management while focusing on research and development efforts. It is essential for them to maintain their existing competitive advantages while continuously expanding upon them to secure a stronger market position. Extending this approach across the entire battery swap ecosystem will necessitate an emphasis on capital management alongside standardized technological development as core elements driving future growth.

4.2 Passenger Vehicles: Exploring User-centric Flexibility

4.2.1 An Overview of the Ecology of the Automotive Battery Swap Industry

Vehicle enterprises are accelerating the launch of battery swapping vehicles and services to expedite the implementation of application scenarios. As the direct stakeholders in the battery swapping mode, vehicle enterprises are also making prompt layout arrangements in the face of the rapid industry development. NEV manufacturers represented by NIO, BAIC, and Geely, as well as traditional vehicle enterprises such as SAIC and GAC AION, are all accelerating the layout of battery swapping services and battery swapping vehicle models. Whether in the passenger vehicle or commercial vehicle fields, the application scenarios are expected to be implemented at an accelerated pace.

NIO is currently a leading enterprise in the battery swapping field for C-end private cars and has launched the BaaS service to reshape the ecosystem model.

Founded in 2014, NIO owns nearly 300 battery swapping patents. In August 2020, the company released the BaaS plan. For users who choose the BaaS mode to purchase NIO cars, there is no need to buy the battery pack when purchasing the car. They can choose to rent battery packs of different capacities according to actual usage needs and pay the service fee on a monthly basis. Meanwhile, the company jointly established a battery asset management company to integrate the resources of capital, operators, and battery manufacturers and carry out battery leasing services. Thus, it has become a powerful promoter of the domestic "vehicle - battery separation" ecosystem.

NIO is accelerating the construction of the battery swapping network, with a vision of planning to build more than 4,000 battery swapping stations. As of December 2024, NIO has filed over 1,400 patents related to battery-swap technology and operates more than 3,400 swap stations across China, forming a highway-based network structured as "nine vertical, nine horizontal corridors covering fourteen urban agglomerations." Together with concurrently deployed super-chargers and destination chargers, the company's overall charging and swapping infrastructure has reached 8,100 sites. Through four generations of technological iteration, the swap duration has been reduced from the approximately 300-second, manually assisted process of the first-generation station introduced in 2017 to a minimum of 144 seconds of unmanned, in-vehicle-automated operation in the fourth-generation station launched in 2024. Each Gen-4 station houses 23 battery slots, six ultra-wide-angle LiDAR sensors, and four Orin-X chips (1,016 TOPS of AI compute), enabling a daily service capacity of 480 vehicles and compatibility with NIO, Onvo (Lèdào), and alliance brand models. Under the 2025 expansion plan, NIO will add roughly 1,600 new swap stations worldwide, lifting the global total beyond 4,000 and establishing a "three-minute energy replenishment" paradigm competitive with—or superior to—conventional refueling.

With the widespread adoption of battery swap technology, an increasing number of automotive enterprises and related companies have entered the realm of battery swap sharing. Consequently, a new business ecosystem for automobiles and energy has emerged, characterized by vehicle-battery separation and battery swapping. As illustrated in the diagram above, this model encompasses seven roles across three industries, in addition to users. Below is an overview of the battery swap sharing business ecosystem categorized by industry.

In the traditional vehicle sales model, batteries are sold by battery manufacturers to vehicle producers as components. These components are then assembled into complete vehicles that are sold together with their respective batteries to end-users. However, within the new battery swap ecosystem, batteries are initially supplied by the battery manufacturer as components to the vehicle producer. When selling a vehicle, manufacturers can offer it without a battery to consumers while simultaneously selling the corresponding battery to a dedicated asset management company specializing in batteries. This particular battery will promptly enter into circulation within the battery swap ecosystem alongside its user's journey through various swaps. Should a battery operation company determine that a specific unit must be decommissioned due to energy depletion or other factors, it will be recycled by a specialized recycling firm; subsequently, its raw materials will flow back to the original manufacturer for reuse.

Automotive Industry: In the emerging battery swap ecosystem, when a vehicle manufacturer sells a vehicle, it can issue two separate invoices: one for the vehicle itself, excluding the battery, to the user; and another invoice for the battery to the battery asset company. Users do not pay for the cost of the battery directly to the vehicle manufacturer at the time of purchase. Furthermore, users will enter into a battery lease agreement with the battery asset company and will be required to pay a monthly rental fee for this service. Throughout the duration of vehicle use, the battery swap operation company will provide necessary battery swapping services to users, who are only responsible for paying standard replenishment fees to this operation company. Although there is a financial separation between the vehicle and its associated battery during purchasing transactions, physically speaking, there is no need for removal of the battery from its designated vehicle.

Energy Power Industry: Based on the characteristics of battery swap stations and the substantial asset nature of the batteries involved, ownership of these swap stations is likely to be distributed rather than solely managed by the battery swap operation companies. Following a light-asset transformation, these companies can more effectively allocate resources toward developing efficient battery swapping technologies. Moreover, due to their inherent advantages in energy storage as well as peak shaving and frequency regulation, energy power companies are expected to play a significant role in the emerging battery swap ecosystem. The latest passenger vehicle battery swap stations available on the market are equipped with integrated rooftop photovoltaic systems and bidirectional high-power charging modules. These systems utilize low-energy consumption smart algorithms to dynamically adjust power distribution, thereby significantly enhancing both charging and discharging efficiency for batteries within the station. This capability facilitates bidirectional interaction and assists the power grid in managing peak loads while filling valleys. Simulation calculations indicate that each station has the potential to save an average of 18,000 kWh annually, contributing positively toward China's goals for carbon neutrality and peak carbon emissions reduction.

In the aforementioned business ecosystem, the battery, automobile, and energy power industries are interconnected while maintaining their operational independence. It is anticipated that through close collaboration among the seven roles within this ecosystem, a high-quality replenishment experience can be delivered to an increasing number of NEV users. This effort aims to contribute significantly to addressing grid acceptance and consumption challenges.

The battery swap model effectively addresses the rapid replenishment needs of NEV users, significantly enhancing battery longevity and safety while lowering the barriers to purchasing NEVs. This approach is particularly advantageous for operators of commercial vehicles who are sensitive to time costs associated with refueling. By leveraging favorable policies, the battery swap model has been widely adopted in operational contexts, notably among taxis.

Operational vehicles are typically custom-designed models characterized by a limited range of brands and standardized battery specifications that align with the technical requirements of the battery swap system. Given the relatively uniform routes taken by B-end models (such as taxis and ride-hailing services), the applicability of battery swapping is broader, facilitating its promotion and implementation; thus far, extensive commercial operations have already been established within this sector. Furthermore, operational vehicles impose stringent demands on both the lifespan and maintenance of power batteries. The battery swap mode allows for trickle charging of these batteries, thereby extending their overall life cycle.

Moreover, operational efficiency is paramount for these vehicles. Fast charging requires a minimum duration of 30 minutes, while conventional fast charging can take between one to one-and-a-half hours; in contrast, battery swapping takes only 3–5 minutes, clearly demonstrating its advantages. As electric vehicle adoption increases, drivers often face challenges locating available fast chargers, which can detract from operational efficiency. By developing an energy service network, companies specializing in battery swaps can substantially enhance this efficiency for operational vehicles while establishing a sustainable business model.

The introduction of centralized battery banks capable of housing power batteries facilitates a vehicle-battery separation sales model that enables drivers to pay solely for components excluding the power batteries themselves. This innovation significantly reduces vehicle acquisition costs and promotes sales growth for vehicles designed with swappable batteries.

Currently, the predominant method for battery swapping in passenger vehicles is undercarriage battery swapping. Mainstream models designed for battery swapping predominantly utilize this approach. The undercarriage battery swapping technology employs specialized devices that, through guiding mechanisms, achieve precise alignment with batteries mounted beneath the vehicle chassis, thereby facilitating efficient removal and installation of the batteries.

In response to the electrification demands within the taxi market, leading traditional original equipment manufacturers (OEMs) have introduced taxi models equipped with battery swap capabilities. These innovations address both the transition from fossil fuels to electric power and meet the high-efficiency recharging requirements prevalent in this sector. Notable examples of mainstream battery-swappable taxi models currently available on the market include BAIC EU300/EU5, SAIC Ei5, GAC Aions, Changan EV460, Dongfeng E70, FAW Hongqi E111/Bestune NAT, JAC IEVA50, Fengxing S50 among others.

4.2.2 NIO's Battery Swap: Practices, Challenges, and Solutions

Currently, among all vehicle enterprises focusing on battery swapping, NIO stands out as a company that has consistently adhered to the battery swap model. The development of NIO's battery swap business for private passenger vehicles is currently in an accelerated growth phase, characterized by intensive expansion and an enhanced energy replenishment experience for consumers.

Distribution and scale: By the end of 2024, NIO had deployed approximately 3,399 battery swap stations, of which 988 were located along highways, covering

550 cities nationwide; cumulative battery swaps exceeded 90 million, with an average of roughly 100,000 swaps performed daily.

Application scenarios: In major urban areas, more than 77% of users have access to at least one swap station within a 3 km radius of their daily activities. Additionally, there are currently 601 highway swap stations deployed along highways that connect six vertical routes, four horizontal routes, and eight major city cluster highway networks.

Operation and business model: The charging standard is set at yuan 50 per day with a maximum cap of yuan 880 for up to 31 days; users are entitled to between four to six complimentary battery swaps each month.

Technical solutions: NIO primarily employs a chassis-bolted battery swap method. Furthermore, by leveraging autonomous driving technology and strategically dense layouts of swap stations, the company significantly enhances the energy replenishment experience for consumers while bolstering brand recognition.

Challenges: Currently, the development of highway swap stations faces key bottlenecks, including limitations in site selection and challenges associated with expanding power capacity.

Development plan: NIO is committed to advancing its highway battery swap network, with an ambitious goal to establish a comprehensive system comprising 9 vertical routes, 9 horizontal routes, and 19 major city cluster highways by 2025. Presently, critical factors influencing the establishment of highway swap stations include site availability and power capacity.

From the perspective of battery swapping positioning, NIO aims to offer diverse charging solutions tailored for private passenger vehicles to accommodate varying charging needs. This approach is grounded in the principle of "charging when convenient and swapping, when necessary," providing four essential service positions for battery swaps catering specifically to private vehicle owners.

Residential charging solutions: The installation of AC and DC charging stations in residential areas enables private passenger vehicle owners to conveniently charge their vehicles at home, effectively addressing their daily charging requirements.

Opportunistic charging: The provision of DC and ultrafast charging stations facilitates convenient and rapid charging during journeys, offering reliable options for long-distance travel as well as temporary stops.

Inter-city travel: Establishing battery swap stations is a crucial strategy for inter-city travel, allowing private passenger vehicle owners to swiftly exchange batteries. This significantly reduces charging time and enhances travel flexibility.

Emergency services: It is essential to ensure that private passenger vehicle owners can promptly meet urgent charging needs, thereby maintaining the convenience and reliability of charging solutions in emergency situations.

The implementation of these long-term strategic initiatives is anticipated to deliver comprehensive, convenient, and efficient charging solutions for private passenger vehicle owners, further advancing the sustainable development of electric vehicles.

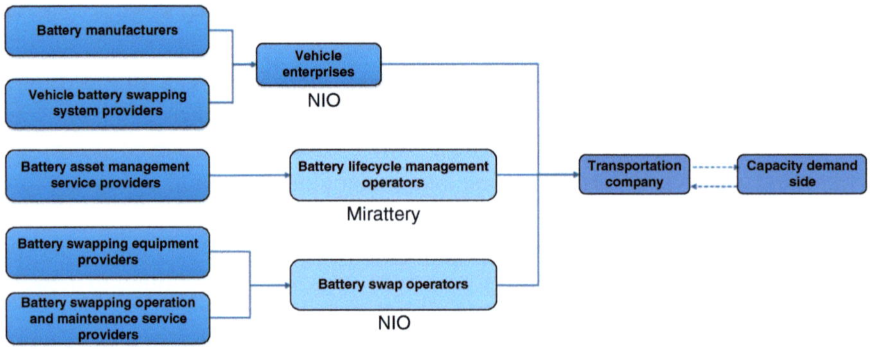

Figure 4.4 NIO's layout in the industrial ecosystem.

These diverse solutions are supported by a broad ecosystem of key stakeholders in the battery swap industry, as mapped out in Figure 4.4.

To ensure the sustainable expansion of the battery-swap ecosystem, two cross-segment bottlenecks must be resolved: (i) the efficiency of market entry for new battery suppliers and (ii) the security of asset transfer in the secondary vehicle market.

1. **The efficiency of market entry for new battery suppliers**
   Whenever a new battery supplier enters the battery swap ecosystem, even if this supplier produces batteries that meet the same standards and specifications, vehicle manufacturers already engaged in the ecosystem are required to re-certify their models compatible with battery swapping. This process incurs significant costs. Assuming there are X models within the ecosystem and Y new battery suppliers are introduced, each model would necessitate certification for every new supplier, leading to a total of XY certifications. Such regulations considerably elevate the barriers for new battery suppliers seeking entry into the market and increase costs for vehicle manufacturers due to redundant certification processes.
   Recommendation: Relevant authorities could categorize batteries of identical specifications and standards as part of a single family. Consequently, vehicle manufacturers would only need to certify their models once for this family of batteries. New battery suppliers, upon receiving certification from the battery swap ecosystem, could then produce batteries according to these established specifications and standards, thereby gaining inclusion in the ecosystem.
2. **The security of asset transfer in the secondary vehicle market**
   Currently, when the original owner transfers a used car with a separated battery to a new owner, there is no requirement to provide proof of battery separation or to wait for the completion of the battery lease agreement transfer. Some original owners intentionally conceal the fact that the vehicle has a separated battery, selling it as if it were an intact used car at full market value. This practice results in financial losses for the battery asset company due to unpaid rental fees from the original owner. Consequently, the new owner experiences dual losses, which fosters negative perceptions in the market regarding resale risks associated with battery-separated vehicles.

Recommendation: It is urged that relevant authorities establish more robust procedures to facilitate efficient and secure transfers of used cars with separate batteries.

Given the differing conditions of batteries in battery swap vehicles at the point of disposal, the existing policy mandating synchronous scrappage for both electric vehicles and their associated batteries may lead to unnecessary waste, particularly concerning those batteries that remain functional.

Recommendation: It is recommended that pertinent authorities consider implementing a system for asynchronous scrappage specifically for battery swap vehicles. This would enable batteries that have yet to meet their disposal criteria to re-enter the battery swap ecosystem for ongoing use, thus prolonging their operational lifespan within society.

4.2.3 Vehicle-battery Separation and Swap Network Development

1. **Practice of Vehicle-battery Separation Model**

   On August 20, 2020, NIO launched its battery as a service (BaaS) program, enabling customers to purchase the electric vehicle chassis separately while leasing the battery. By physically decoupling the vehicle from the battery, NIO effectively separates the cost of the battery from that of the vehicle itself. Under this BaaS model, users benefit from a reduced initial purchase price for the vehicle and pay a monthly lease fee for the battery by subscribing to the BaaS service.

   To establish a complete BaaS service, on August 18, 2020, NIO collaborated with Contemporary. To establish a comprehensive BaaS offering, on August 18, 2020, NIO partnered with Contemporary Amperex Technology Co., Ltd. (CATL), Hubei Science and Technology Investment Group Co., Ltd., and a subsidiary of Guotai Junan International Holdings Limited to invest in Wuhan Weineng Battery Asset Co., Ltd. (hereinafter referred to as "Wuhan Weineng"), with NIO holding approximately 19.8% equity in Wuhan Weineng.

   Under the BaaS model:

   1. In the transaction between NIO and Wuhan Weineng, NIO supplies batteries to Wuhan Weineng as part of its battery asset company operations. The commitment to transfer control of the battery pack constitutes NIO's sole performance obligation under the sales contract for the battery pack with the battery asset company. From a financial perspective, upon delivery of the vehicle (including the battery pack) to the BaaS user, NIO recognizes revenue from selling the battery pack to Wuhan Weineng, at which point control of the battery pack is transferred to Wuhan Weineng.
   2. Between the user, NIO, and Wuhan Weineng, NIO sells the vehicle to the BaaS user, with both parties entering into a Vehicle Purchase Agreement. The user leases the battery from Wuhan Weineng and pays a lease fee; concurrently, the user and Wuhan Weineng sign a Power Battery Lease and Service Agreement. This arrangement stipulates that ownership of the battery resides with Wuhan Weineng while granting the user rights to utilize the battery through leasing.

Furthermore, when selling battery packs, NIO has established a service agreement with Wuhan Weineng. Under this agreement, NIO provides various services to Wuhan Weineng including monitoring of battery packs, maintenance, upgrades, replacements, and IT system support, among others. In return for these services, NIO charges Wuhan Weineng a monthly service fee. Should the user fail to remit payment for their battery lease fees, Wuhan Weineng reserves the right to request that NIO track and lock the user's battery in order to restrict its usage. Additionally, in an effort to foster BaaS development further, NIO offers guarantees to Wuhan Weineng aimed at mitigating risks associated with users defaulting on their rental payments; however, such guarantees shall not exceed the total service fees received by NIO from Wuhan Weineng.

As advancements in battery technology occur over time, users are afforded flexibility in choosing either an upgrade to a higher-capacity battery or switching models altogether. Within this BaaS framework—wherein ownership of initially installed batteries is retained by the asset company—when users express interest in upgrading their batteries, it is effectively an upgrade of assets owned by said company facilitated by NIO. Post-upgrade implementation will result in BaaS users incurring increased lease fees payable to the asset company for renting larger-capacity batteries.

2. **Practice in Battery Swap Network Construction**

Since the introduction of NIO's proprietary electric vehicle battery swap technology in 2018, more than 2,400 battery swap stations have been established as of June 13, 2024. These stations provide efficient and safe energy replenishment services for over 500,000 pure electric vehicles. With the maturation of the battery swap network, NIO launched a flexible battery upgrade service in April 2021 to cater to individualized battery usage needs. Concurrently, a closed-loop battery recycling system has been developed based on the extensive coverage of this network.

Currently, NIO's high-speed battery swap network in China encompasses seven north–south expressway corridors: G1 Jingha Expressway, G2 Jinghu Expressway, G3 Jingtai Expressway, G4 Jinggangao Expressway, G5 Jingkun Expressway, G15 Shenhai Expressway, and the Guangzhou section of the G45 Daguang Expressway. Additionally, it includes six east–west expressways: G20 Qingyin Expressway; Suzhou-Shaanxi section of the G30 Lianhuo Expressway; G40 Hushan Expressway; G42 Hurong Expressway; G50 Huyu Expressway; and Shanghai-Hunan section of the G60 Shenhui Expressway.

Moreover, this comprehensive network spans across 11 metropolitan areas, including the Beijing-Tianjin-Hebei region; Yangtze River Delta; Greater Bay Area; Chengdu-Chongqing area; Shandong Peninsula; Mid-Yangtze River region; Guanzhong area; Central Plains region; Central Shanxi province; Beibu Gulf area, and Western Taiwan Strait. This results in a robust high-speed battery swap infrastructure comprising seven north–south corridors and six east–west corridors that collectively serve these metropolitan regions.

The network not only covers 675 highway service areas and connects with 125 highway exits nationwide but also links over 700 cities across China. As of

Figure 4.5 Map of NIO's nationwide battery swap station layout.

May 2024, these swap stations have facilitated an average of over 60,000 daily swaps for users—culminating in more than 46 million total swaps—thereby creating a fast and convenient energy replenishment service that enhances the overall driving experience. The specific scale and composition of this charging and swapping network layout are shown in Figure 4.5.

There are 3,539 battery swapping stations, among which 1,002 are located on expressways.

Expressway Battery Swapping Network

There are 8,307 charging stations with 27,050 charging piles. Forming the Power Journeys charging landscape. More than 1.326 million third-party charging piles have been connected as of 26 October 2025.

As of May 2025, NIO has established a cumulative total of 60 battery-swap stations and 19 charging stations (33 charging connectors) across five European countries—Norway, Germany, the Netherlands, Sweden, and Denmark—and has integrated access to more than 600,000 third-party public charging points.

From 2007 to 2012, it was the early stage of domestic battery swap stations, primarily serving buses, heavy trucks, taxis, and municipal vehicles. At this time, the swap stations were generally building-type, using large workshops to achieve battery swaps, occupying an area of around 1,000 square meters. Figures 4.6–4.8 show these early stage battery swap stations.

The land, electricity, and labor requirements associated with traditional building-type battery swap stations have led to prohibitively high costs. This has restricted their deployment to highly centralized locations, which is not conducive to the growth of industry, particularly for private vehicle battery swaps. Consequently, after 2013, the battery swap business in China entered a dormant phase.

Beginning in 2016, however, the rapid advancement of the domestic electric vehicle industry created an urgent demand for enhanced energy replenishment and

NIO Power
Europe
Dombås
Alvdal
Oslo, Furuset
Power Swap Station NO. 56
Power Charger Station NO. 22
Lillehammer
NORWAY
SWEDEN
Tangen
Gol
Ringerike, Valdresporten
Straume
Bryn
Kløfta
+2
Lier
Vestby
Enköping
Arlandastad
Haugesund
Seljord
Tønsberg
Sarpsborg
Stockholm, Frihamnen
Stavanger
Bamble
Mariestad
Mjölby
Kristiansand
Kungälv
Varberg
DENMARK
Malmö
Slagelse
Assen
Power Swap Station NO. 55
+1
NETHERLANDS
Utrecht
Apeldoorn
Amsterdam
Harmelen
Den Hoorn
Dordrecht
Tilburg
Emsbüren
Arnhem
Kirchlengern
Großburgwedel
Lehre
Berlin, Spandau
Dorsten
Hilden
Schipkau
Schkeuditz
Maastricht
Reiskirchen
GERMANY
Sulzbach
Waldlaubersheim
Geiselwind
Aurach
Regensburg
Ulm Seligweiler
Zusmarshausen
Mühldorf
56 Power Swap Stations
22 Power Charger Stations
600,000+ accessible 3rd party chargers
*As of 5th September, 2024 in Europe

Figure 4.6 The 2008 Beijing Olympic Games battery swap station, completed in July 2008, served 50 electric buses using a single-lane dual-parking space format, marking the first implementation of battery swap services for purely electric vehicles in China.

Figure 4.7 The 2010 Shanghai World Expo battery swap station, completed in May 2010, served 120 buses with an access power of 8 MW.

improved infrastructure efficiency. This shift prompted further maturation of battery swap technology. Battery swap stations designed for taxis and private vehicles began to be deployed at an accelerated pace. The new generation of swap stations predominantly adopted preassembled models that utilize modular solutions undergoing systematic verification processes; these can be swiftly installed on-site.

These preassembled battery swap stations are significantly optimized in terms of size and require less land area as well as reduced electrical capacity. As a result, they lower overall holding costs and are more suitable for distributed deployment, thereby reaching a broader user base. Furthermore, thorough preinstallation verification enhances equipment reliability. Preassembled swap stations have thus emerged as the mainstream format for contemporary battery swapping operations.

Table 4.3 presents illustrative case of pre-installed passenger car transfer stations.

Figure 4.8 The State Grid Xuejia Island bus battery swap station, completed in July 2011, served 200 electric buses with an access power of 10 MW.

Table 4.3 Comparison of typical preassembled passenger vehicle swap stations.

Swap Station	NIO Gen 2	NIO Gen 3	Geely L Cube	Aulton	AnYixing	CATL EVOGO
Application Scenario	Private Car Swap	Private Car Swap	Taxi Swap	Taxi Swap	Taxi Swap	Taxi Swap
Equipment Footprint	Approx. 60 sqm	Approx. 70 sqm	Approx. 120 sqm	Approx. 130 sqm	Approx. 90 sqm	Approx. 60 sqm
Electrical capacity	550 kVA	630 kVA	630 kVA	2400 kVA	550 kVA	630 kVA
Layout form	Single Swap Bay + Single Battery Bay	Single Swap Bay + Single Battery Bay	Single Swap Bay + Dual Battery Bays	Single Swap Bay + Dual Battery Bays	Single Swap Bay + Single Battery Bay	Single Swap Bay + Single Battery Bay

In the early stages of battery swapping, human intervention was common in many processes. Reasons for human involvement varied, including:

- insufficient vehicle design compatibility, such as gear shifting methods and battery pack movement paths that did not match well with battery swapping;
- lack of complete and reliable automation solutions, such as battery transportation and installation;
- low accuracy in device recognition and movement, such as poor positioning ability of battery installation equipment requiring manual assistance;
- choosing manual methods to reduce costs when the deployment volume was insufficient.

Figure 4.9 Hangzhou taxi project in 2012 used semi-automatic battery swapping.

Examples of such early-stage manual and semiautomatic battery swapping operations are depicted in Figure 4.9.

Currently, fully automated solutions for battery swap equipment are basically mature in the industry, and the reliability of newly deployed products has significantly improved.

The main factors affecting unmanned operation of swap stations include:

- reliability of critical components (especially moving parts): such as battery locking mechanisms, battery transportation mechanisms, and vehicle lifting mechanisms;
- reliability of cloud software interconnection: including anti-foolishness design of cloud platforms, success rate, and stability of station interconnections;
- impact of extreme weather conditions: particularly icy conditions under the chassis during winter rain and snow;
- reliability of fire protection systems.

Taking NIO as a case study, significant advancements have been achieved in unmanned operations. As of September 2023, the duration of unmanned operation at NIO's Gen 2 and Gen 3 swap stations reached approximately 90%, with manual intervention rates limited to only 2–3%. Most faults can be addressed remotely. Notably, the development of automation has markedly enhanced the speed of battery swapping, thereby improving operational efficiency and user experience.

Through optimization of battery locking mechanisms, swapping procedures, and battery handling structures, leading battery swap service providers have realized substantial improvements in swapping times. For instance, NIO employs a bolted structure; the swapping time for its first-generation station was around seven minutes. In contrast, actual operational data for the second and third generations indicates an average time of approximately three minutes, with future products anticipated to optimize this further to under two minutes. Aulton New Energy utilizes a latch mechanism; its earlier models had a swapping time close to five minutes, while newer products are capable of achieving a remarkable speed of just 1.5 minutes.

Currently, the average daily swap orders for taxi operations typically exceed 150 orders per day; some companies even surpass 500 orders under peak conditions.

With the advancement of intelligent networking technology and the enhancement of hardware and software capabilities at battery swap stations, these facilities have transformed from mere sites for battery charging and old-to-new battery replacement into comprehensive service platforms. These platforms now integrate intelligent dispatching, autonomous driving, big data collection, battery monitoring, battery life prediction and annual inspection, smart settlement processes, as well as energy trading functionalities. Below are some typical examples:

4.2.3.1 Driving Assistance

Early battery swap operations primarily targeted commercial vehicles and taxis, where drivers typically underwent rigorous training and possessed extensive long-distance driving experience. The success of the parking process during a battery swap heavily depended on the skill level of the driver.

As private vehicle battery swap services emerged, ensuring that ordinary drivers could safely complete the parking process became essential for enhancing the overall operational efficiency of these services. Driving assistance functions that leverage collaboration between vehicles and stations have shown significant advantages in this context.

For instance, NIO began developing automatic parking features specifically designed for swap stations with its second-generation models. This innovation replaced manual parking by enabling autonomous driving capabilities within the vehicle, thereby eliminating the need for personnel to park it. Battery swap parking facilitated by driving assistance offers three primary advantages:

Enhanced parking success rates, minimizing the risk of station collisions, and ensuring the safety of both individuals and vehicles.

Improved user experience, as the entire battery swapping process can be completed without requiring users to exit their vehicles, thereby providing a superior experience, particularly in extreme weather conditions. More consistent parking durations contribute to operational efficiency.

4.2.3.2 Battery Monitoring

Safety concerns associated with power batteries, including self-discharge, insulation failure, internal short-circuiting, overcharging, and over-discharging, can result in thermal runaway of the battery pack. This phenomenon poses a significant risk of fires and explosions affecting multiple batteries or even the entire swap station. According to Heinrich's Law, for every serious accident that occurs, there are typically 29 minor incidents and 300 near-misses, along with approximately 1,000 precursors to accidents.

As a facility designed for the centralized storage of power batteries, swap stations must be equipped with comprehensive capabilities for full-process battery monitoring, early warning systems, and fire protection measures to safeguard personnel and property associated with the station. The safety of batteries within the context of swap stations is primarily influenced by several key factors:

Intrinsic factors: The level of research and design dedicated to battery technology as well as the stability of manufacturing processes.

Extrinsic factors: improper use of charging equipment, abnormal logic during the charging process, physical damage, or water ingress to the battery during vehicle operation.

For instance, NIO enhances the safety and reliability of its vehicles and batteries throughout their lifecycle by implementing a cloud-coordinated battery monitoring system. This system is capable of real-time and offline monitoring of battery packs at the swap station. The cloud platform leverages deep learning algorithms to analyze operational data from batteries as well as historical fault cases, allowing for the early detection of abnormalities before they are noticed by users. It proactively identifies potential fault types and facilitates preventive measures along with immediate on-site risk mitigation.

NIO's battery monitoring system at the swap station encompasses the following aspects:

1. Structural damage assessment: This involves identifying any damage to the top and bottom covers of the battery during each entry and exit from the swap station, conducting regular structural inspections on stored batteries, and tracing any damage incurred during transportation or storage.
2. Battery management system (BMS) parameter monitoring: This includes detecting instances of overcharging or overdischarging, daily state of charge (SOC) assessments, as well as continuous monitoring of voltage, current, internal resistance, insulation status, and overall battery consistency.
3. Battery safety status prediction: Developing a Comprehensive Big Data Model for Battery Safety Utilizing Battery Data and Deep Learning Algorithms to Facilitate Early Detection of Anomalies.
4. Emergency response for abnormal batteries: Executing Emergency Fire Interventions through BMS Information and Monitoring of the Battery Bay When the Battery is Identified as Being at Risk.

4.2.3.3 Off-peak Charging

Electricity costs represent a fundamental expense in the operation of swap stations. In China, a tiered pricing system is typically employed, characterized by relatively stable prices over extended periods. Conversely, electricity prices in Europe display significant dynamic trading characteristics, with hourly price fluctuations that can even include negative pricing at times, alongside daily variations.

Consequently, leveraging periods of lower electricity costs is essential for both domestic and international operations. Off-peak charging necessitates an effective balance between electricity pricing information and battery swap service data. This process relies on sophisticated algorithms to optimize the coordination of battery swapping services and charging activities, thereby maximizing the utilization of low-cost periods to substantially reduce operational expenses.

4.2.3.4 Vehicle/Station Grid Interaction

Vehicle-grid interaction refers to the exchange of energy and information between electric vehicles and the power grid. From an energy flow perspective, this interaction can be categorized into two types: one-way interaction, characterized by orderly charging with controlled charging power, and two-way interaction, which involves bi-directional charging and discharging. Electric vehicles and battery swap stations possess remarkable regulation capabilities along with significant potential for load adjustment. If these resources are utilized in accordance with the operational requirements of the power system, timely and appropriate power interactions can facilitate the balancing of peak and off-peak loads, alleviate congestion, and support real-time power balance as well as enhance grid safety, economic efficiency, and reliable electricity supply.

At scale, vehicle/station–grid interaction evolves into a commercial activity wherein resource aggregators coordinate distributed swap stations alongside electric vehicle users through both commercial strategies and technical means. This approach aims to fully leverage the adjustable load potential of swap stations while creating tradable commodities related to power and energy that can be sold to entities requiring flexible resources.

4.2.3.5 The Primary Business Model and Technical Foundation Includes

From the perspective of a local grid, orderly charging and discharging can significantly enhance the utilization rate of the distribution network and improve economic efficiency.

From the standpoint of the distribution network, orderly charging and discharging not only optimize the use of distribution network assets but also bolster economic efficiency. Furthermore, this approach can be extended to integrated light-charge-swap stations, thereby promoting local consumption of photovoltaic power. The term "orderly charging" refers to selecting times for charging when there is available capacity in the grid.

For instance, as illustrated in Figure 4.10, implementing orderly charging within the distribution network can significantly lower the electricity costs associated with charging at swap stations. Additionally, aggregators have the opportunity to generate revenue by offering ancillary power services.

Bidirectional swap stations are capable of participating in local (behind-the-meter) energy services, thereby enhancing revenue streams through various means such as power balancing, peak shaving, and valley filling, as well as promoting local consumption of renewable energy. This approach not only expands new business scenarios but also increases overall revenue, broadens cooperative partnerships, and improves locational opportunities. The conceptual framework for such participation in regional grid services is shown in Figure 4.11, while a specific technical implementation framework, as exemplified by NIO, is detailed in Figure 4.12.

On the basis of the above scenarios, aggregating more swap stations can further participate in frequency regulation and peak shaving of a wider area grid (such as a

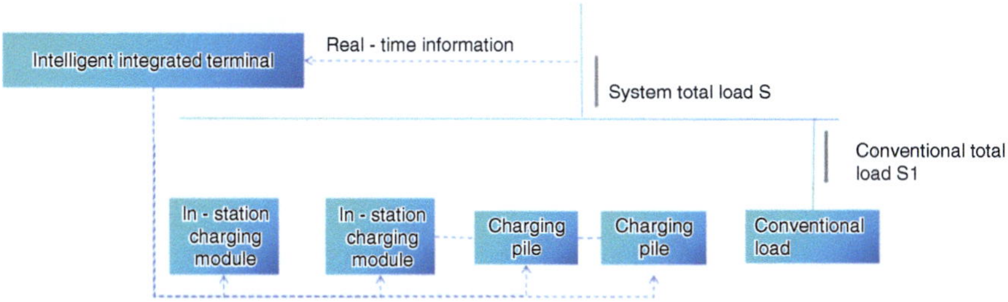

Figure 4.10 Participation of swap stations in wide-area grid (regional grid) frequency regulation and peak shaving or demand-side management.

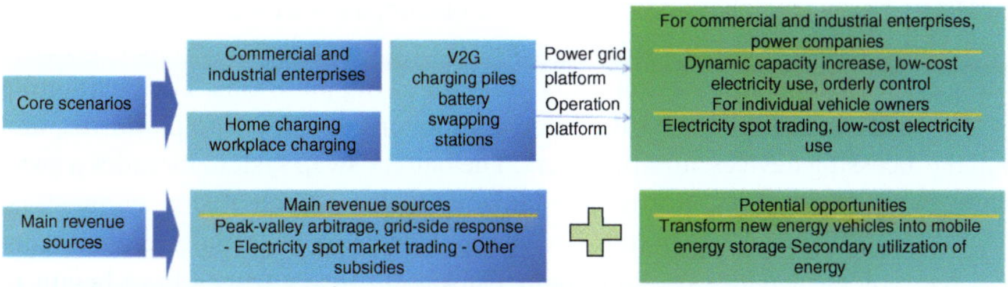

Figure 4.11 A comprehensive framework of electric vehicle-to-grid (V2G) business models and value streams.

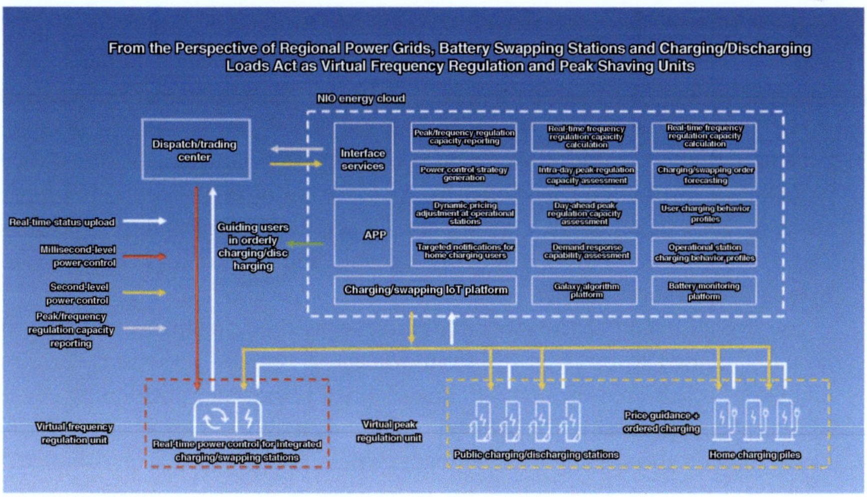

Figure 4.12 Framework for NIO's participation in regional grid frequency regulation and peak shaving.

provincial grid). This is the most challenging application scenario, involving many stakeholders and complex technology, but once accomplished, it would be the most significant and require long-term planning.

4.3 Commercial Vehicles: Pathways to Cost Reduction Through Scale

4.3.1 An Overview of the Battery Swapping Model for Heavy-duty Trucks

From the previous chapter, we have learned that the commercialization of battery swapping for commercial vehicles is predominantly concentrated in the Chinese market. In China, research, development, application, and practical implementation of related technologies are at the forefront on a global scale. The following main sections will focus on the battery swap industry within China.

Electric heavy truck battery swap stations encompass various systems including positioning systems, battery swap systems, operation and maintenance systems, safety systems, logistics systems, and data monitoring platforms. The positioning system consists of vehicle identification components, positioning platforms, and signal processing devices, among others. The battery swap system includes a swap platform, stacker cranes, lifting mechanisms, locking mechanisms, connectors, etc. Currently, underbody battery swapping represents a partially commercialized technological pathway; some light trucks and heavy trucks have begun to adopt this technology. Electric medium and heavy trucks primarily utilize three types of battery swapping technologies: top-lift swap, whole-side swap, and dual-side swap.

1. **Top-lift Battery Swap Method**

   The top-lift battery swapping system primarily employs steel cables to elevate the battery pack, retrieving the battery box designated for replacement from the upper section of an awaiting heavy truck and hoisting it into the designated battery storage area. Subsequently, a positioning and locking device is installed, facilitating the successful replacement of the battery box on the truck that requires a swap. Conventional workshops utilize overhead cranes equipped with two to three degrees of freedom and specialized lifting tools to manage the battery system.

   The overall height of the swap station measures 6 m, featuring a V-shaped groove designed to position the front wheels of vehicles accurately. Precise alignment is achieved through guiding inclined planes and positioning cones located at both the base of the battery system and on the vehicle's swap platform. Depending on the spacing between battery slots, each swap can be completed within 3–5 minutes. Each station accommodates seven spare batteries and has a capacity to facilitate 150–200 swaps.

   Due to limitations in equipment precision for vehicle positioning, occasional significant deviations may occur during swaps, necessitating intervention by onsite personnel to stabilize this process. The onboard swapping mechanism incorporates a six-point pneumatic pin locking system that compensates for flatness discrepancies between the bottom surface of both the battery system and the swap base using flexible pads. This design offers advantages such as simplicity in structure and cost-effectiveness.

Given its straightforward technology, low costs, and practical feasibility requirements—whereby lifting tools must be positioned above vehicles—the overall height of this equipment tends to be relatively elevated; this can lead to temporary building approval challenges in certain urban construction projects. The top-lift battery swapping solution represents one of the earliest commercial implementations available today and remains one of the most widely adopted solutions for heavy truck swaps in current markets, as illustrated in Figure 4.13. Notable companies involved include Shanghai Enneagon Energy Technology Co., Ltd.

2. **Side-swap Battery Swap Solution**
The side-swap battery swap solution primarily encompasses two methods: the side-grip side-swap and the top-grip side-swap. By employing a LiDAR recognition system, the precise position of both the vehicle and its onboard battery system is determined with high accuracy. Multi-degree-of-freedom battery swap robots, along with intelligent telescopic robotic arms, execute the battery swapping task by gripping and maneuvering the battery system from either the side or top.

Figure 4.13 Top-mounted battery swap mode.

The overall height of the swap station is approximately 4.5 mm, occupying an area of 200 square meters. The average time required for a swap is between three and four minutes, enabling service for over 300 vehicles while maintaining a remarkable swap success rate of 100%. The swap station integrates advanced technologies such as 5-axis robots, LiDAR systems, machine vision, and other equipment to effectively coordinate each swapping operation.

This method is suitable for various types of vehicles in urban settings, such as dump trucks, tractors, and concrete mixers, and has relatively low requirements for how drivers park for battery swaps. The intelligent equipment also adds significant value. Moreover, the advantage of the single-side swap method is that the main equipment height of the swap station is similar to the vehicle height, making it easier to classify as equipment rather than a temporary structure during city construction, thus avoiding the temporary building approval process. Single-side whole-body swaps feature high precision, good adaptability, high levels of automation, and the capability for unattended operation, although the initial cost is relatively high. As shown in Figures 4.14 and 4.15, representative companies include ZL IoT Technology Co., Ltd.

Depending on user specifications, the number of available battery slots can be configured to range from 4 to 16. The onboard swapping mechanism utilizes eight pneumatic claws designed to securely grip the bottom of the battery system. This design incorporates a milling technique that flattens the entire surface area to ensure tight contact; this guarantees structural stiffness under locked conditions and prevents impacts or bouncing caused by clearance during vibrations.

This technological approach provides significant advantages in terms of stiffness, achieving a first-order frequency exceeding 17 Hz. Furthermore, it features a locking mechanism characterized by high reliability and durability.

Figure 4.14 Side-grip and side-swap type.

Figure 4.15 Top-grip and side-swap type.

3. **Dual-Side Whole Body Swap**
Compared to the single-side whole body swap technology, the dual-side whole body swap utilizes largely similar technological principles, with a key distinction being the implementation of two sets of equipment that facilitate simultaneous battery swaps on both sides of the vehicle. Dual-side swaps incorporate highly redundant and integrated multi-degree-of-freedom battery swap robots, robust composite intelligent telescopic mechanical arms, as well as advanced multi-sensor fusion technologies such as machine vision target recognition, laser ranging, and encoder positioning to execute the swapping process efficiently.

In specific scenarios where vehicle weight is substantial and extending the wheelbase for battery placement behind the cab is not feasible, coupled with constraints imposed by unique terrain conditions, batteries must be positioned on both sides of the vehicle. This approach incurs costs that are nearly double those associated with single-side swaps; however, it plays an indispensable role in delivering swapping services under particular circumstances.

Currently, dual-side swap technology finds its primary application in mining trucks. As illustrated in Figure 4.16, State Grid Electric Vehicle Service Co., Ltd. serves as a representative company within this sector.

4. **Underbody Battery Swap**
Underbody battery swapping represents a significant technological advancement for electric heavy trucks. By positioning batteries beneath the chassis, this approach provides several advantages, including a lower center of gravity, enhanced driving stability, unobstructed rear visibility for the driver, maximized cargo space in dump trucks, and alignment with the driving habits of truck operators.

Figure 4.16 Dual-side whole body swap.

However, underbody battery swaps necessitate advanced engineering due to their complexity and stringent reliability requirements associated with locking connectors and swap technology. These demands surpass those found in rear-mounted swaps, resulting in a slower development process compared to their rear-mounted counterparts. Currently, numerous OEMs are actively researching underbody battery-swapping systems for heavy trucks. As illustrated in Figure 4.17, companies such as Yutong Heavy Trucks, alongside new entrants like DeepWay, have successfully launched products into the market.

Underbody battery swap stations typically feature either single-side or dual-side configurations. Single-side swaps offer benefits such as reduced costs and smaller footprints; conversely, dual-side swaps provide greater efficiency and an improved user experience during the swapping process.

5. **Integrated Battery-motor Swap**

Integrated battery-motor swapping represents an innovative battery exchange model developed by the US startup Revoy, specifically tailored for the American market, which boasts a substantial fleet of existing diesel-powered heavy trucks. This approach eliminates the need to replace or modify current trucks; instead, it integrates both the drive motor and battery into a single unit that functions as a self-contained electric propulsion module. This module is positioned between the head tractor and the rear trailer, creating a novel swapping method akin to hybrid propulsion.

The primary advantage of this system lies in its compatibility with the US tractor market, facilitating the conversion of heavy trucks into NEVs without necessitating vehicle replacement. This characteristic promotes large-scale adoption and application. However, one notable drawback is that the cost associated with each swap system may be relatively high, placing financial burdens on drivers who would need to cover expenses for both the battery and the complete drive system. Furthermore, alterations in vehicle length could present regulatory challenges across various countries and regions.

Figure 4.17 Underbody swap.

Currently, this swapping product remains in its early stages of market promotion; thus, its adaptability to market demands and potential establishment as a free swap system will require further validation through real-world testing.

4.3.1.1 Introduction to the Battery Swap Process

Vehicle positioning: The electric heavy truck enters the swap station and is positioned using an intelligent guidance system to ensure accurate parking within the swap area.

Battery identification and gripping: The swap robot automatically locates and identifies the battery position using visual recognition and laser sensors, then grips the battery using a bi-directional mechanical arm.

Battery swapping: The robot removes the battery from the vehicle and installs a fully charged battery from the storage area onto the vehicle.

Battery locking: After the battery is installed, it is securely fastened to the vehicle using high preload and high tolerance locking technology.

Completion of swap: Upon completion of the swap, the vehicle exits the station, ready to return to service.

Figure 4.18 provides a simulation diagram summarizing the key stages: vehicle entry, battery unlocking, robotic swapping, and wheel alignment.

As battery swap technology for electric heavy trucks continues to advance, the operational models are also evolving. Initially, private ownership and financial leasing were the predominant approaches. However, as the ecosystem becomes more defined, electric heavy truck battery swap services are progressively transitioning toward a public service network. Currently, four operational models for electric heavy truck battery swaps have been established, as illustrated in Table 4.4.

The economic benefits of battery swapping for heavy-duty trucks are becoming apparent, with the total lifecycle cost being 16% lower than that of diesel-powered trucks. Given the diesel price of 7.61 yuan (as of January 10, 2023, a low point in

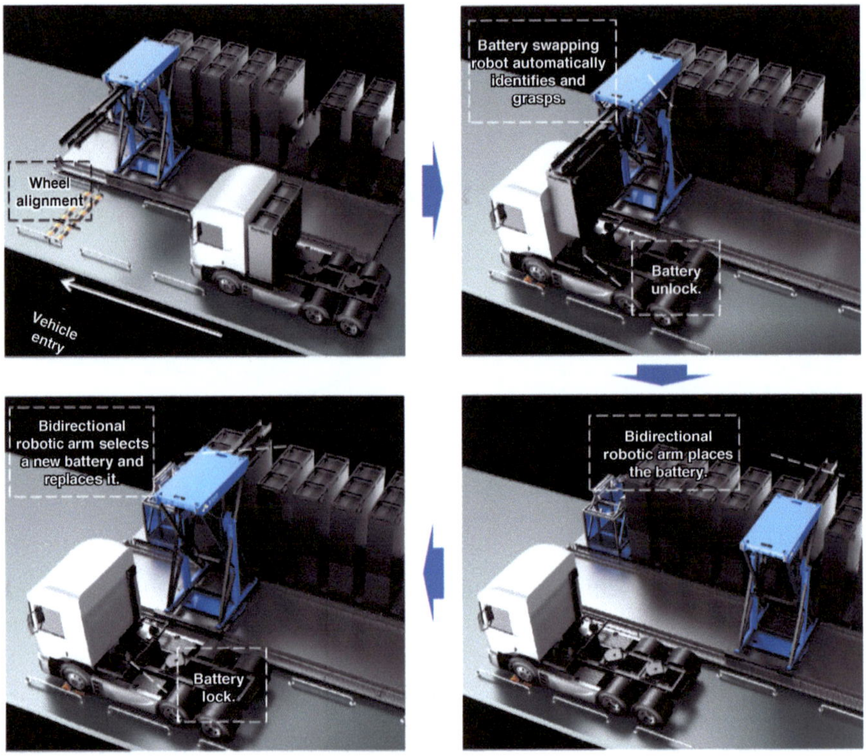

Figure 4.18 Simulation diagram of swap process. *Source:* Zhiliwulian.

the 1–2 month price range) and a battery swapping service fee of 11 yuan/kWh, our calculations show that the initial purchase cost of a battery swapping heavy-duty truck (including the battery) is 84% higher than that of a diesel truck. However, the annual energy expenditure is 39% lower than the diesel fuel costs. Assuming a usage period of 5 years, the total lifecycle cost of a battery swapping truck is nearly 16% lower than that of a diesel truck. Considering the future decrease in battery costs and the corresponding decrease in battery leasing costs, if both battery costs and leasing fees decrease by 10%, the total lifecycle cost reduction for battery swapping trucks will increase by an additional 1.5 percentage points. Given that the economic benefits of battery swapping trucks are negatively correlated with diesel prices, we have calculated the oil price equilibrium point where the costs of diesel trucks and battery swapping trucks are equal to be 5.99 yuan. Since March 2021, diesel prices have generally been above this equilibrium point. As battery costs decrease in the future, the oil price equilibrium point will continue to decrease.

In the future, there exists significant potential for a business model that fosters collaboration among infrastructure investors, swap service operators, battery asset investors, and asset management service providers specializing in battery monitoring. This collaborative approach will offer more comprehensive services and support for electric heavy truck battery swaps, thereby effectively promoting the advancement of the electric heavy truck battery swap industry.

Table 4.4 Analysis of the total lifecycle cost of different types of trucks.

Project	Conventional fuel truck	Electric truck (separate vehicle and battery)	Electric truck (including battery)
Initial vehicle cost (10,000 yuan)	40	45	80
Purchase tax (10,000 yuan)	3.54		
Initial purchase cost (10,000 yuan)	43.54	45.00	80.00
Initial purchase cost increase		3.35%	83.74%
Unit energy consumption (L/kmkWh/km)	0.4	1.7	1.7
Unit energy price (yuan/L, yuan/kWh)	7.61	1.1	1.1
Annual operating days	350.00	350.00	350.00
Daily operating distance (km)	400.00	400.00	400.00
Annual operating distance (km)	140,000	140,000	140,000
Annual energy cost (10,000 yuan/year)	42.62	26.18	26.18
Annual battery rental cost (10,000 yuan/year)		8.3	
Annual subsidy cost (10,000 yuan/year)	42.62	34.48	26.18
Subsidy cost Reduction		−19.09%	−38.57%
Maintenance cost (10,000 yuan/year)	4.6	3.6	3.6
Annual operating cost (10,000 yuan/year)	47.22	38.08	29.78
Depreciation period (Years)	5	5	5
Total lifecycle cost (10,000 yuan) (discount rate 4.88%)	248.64	210.41	209.36
Total lifecycle cost reduction	40	−15.37¥	−15.80¥

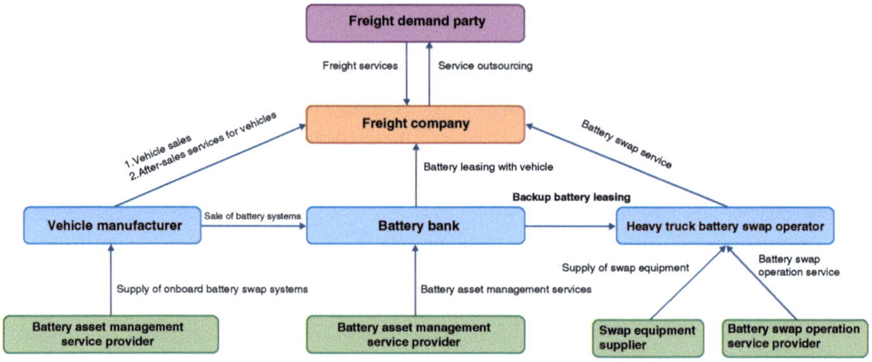

Figure 4.19 Schematic diagram of vehicle-battery separation model.

Among these considerations, it is crucial to emphasize the separation of vehicle and battery models, as illustrated in Figure 4.19. This electrification model for heavy trucks allows consumers to purchase only the vehicle while leasing the battery. In this framework, leasing the battery functions similarly to purchasing fuel, enabling optimized cost control and utilization. By adopting this model, the initial acquisition cost of all-electric heavy trucks is significantly reduced; additionally, it enhances the residual value of the vehicle chassis and improves user acceptance and experience. Consequently, this approach has great potential to address challenges associated with heavy truck electrification. It presents a viable solution for advancing electrification in heavy trucks and is anticipated to see broader application in the future.

4.3.2 The Market for Heavy-duty Truck Battery Swapping and Mainstream Technologies

Battery-swappable commercial vehicles in China began to hit the market from 2020 onwards. As shown in Table 4.5, eight battery-swappable commercial vehicle models were announced in 2020. By 2023, the number of announced models had reached 346, with a cumulative total of 807 models announced over the four years. Among these, 90% of the announced models are battery-swappable heavy trucks, light trucks account for 9%, and the remaining 1% are other types of vehicles.

In 2022, Chinese automakers began to announce and launch a significant number of light truck models, marking a year later than their heavy truck counterparts. As illustrated in Table 4.5, prior to 2022, various manufacturers were experimenting with battery swapping technology for light trucks; however, major players had only introduced one model each. Following at least a year of market feedback and technological advancements, leading automakers accelerated their efforts and increased the volume of new releases in 2023. JAC Motors, Geely New Energy, and King Long United emerged as pioneers in the realm of battery-swappable light trucks. The rankings and detailed announcements of these leading automakers are presented in Table 4.6.

2020 marked the onset of significant growth for new energy heavy trucks in China. As illustrated in Figure 4.20, driven by a combination of industrial policies

Table 4.5 The number of various categories of battery swap-enabled commercial vehicle models announced in China from 2020 to 2023.[a]

Type	2020	2021	2022	2023	Cumulative
Micro truck	0	0	0	6	6
Light truck	3	6	19	45	73
Medium truck	0	0	1	2	3
Heavy truck	5	156	271	293	725
Total	8	162	291	346	807

a. Truck classification includes trucks and special vehicles, according to the carrying tonnage of the truck can be divided into micro-card (total mass < 1.8 tons), light truck (1.8 tons < total mass ≤ 4.5 tons), medium truck (4.5 tons < total mass ≤ 12 tons), heavy truck (total mass > 12 tons).

Table 4.6 Top 10 automakers by total number of announced light truck models in China.[a]

Brand	2020	2021	2022	2023	Total
JAC	1	0	0	8	9
Geely New Energy	0	1	1	4	6
King Long	0	0	0	5	5
Chengdu Dayun	0	0	1	3	4
Dongfeng Motor	0	0	0	4	4
Dongfeng Special Vehicles	0	0	1	3	4
Sinotruk Chengdu Ace	0	0	0	3	3
Beiqi Foton	0	0	0	2	2
Guangxi Automobile	1	0	1	0	2
Hubei Thi-Ring	0	0	1	1	2
Other	1	5	14	12	32
Cumulative	3	6	19	45	73

a. The brand name is short for automobile enterprise.

and market demand, the annual compound growth rate of new energy heavy trucks from 2020 to 2023 reached an impressive 287%. This surge propelled annual sales to 31,034 units in 2023, with pure electric heavy trucks constituting 90% of total sales.

Simultaneously, battery-swappable heavy trucks have increasingly emerged as a vital segment within the pure electric truck category. From 2022 to 2023, the market share of battery-swappable heavy trucks rose significantly—from 35.67% in 2021 to 55.05% in 2022 and subsequently stabilizing at approximately 47.37% in 2023—thereby capturing a substantial portion of the new energy heavy truck market.

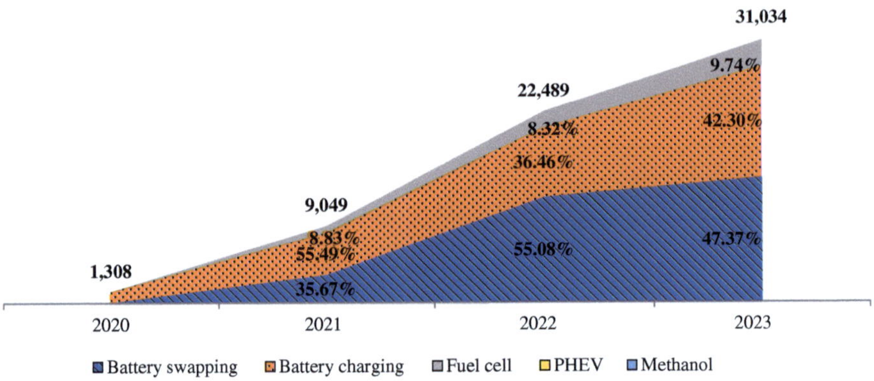

Figure 4.20 Sales volume and share of new energy heavy trucks in China by year (units, %).

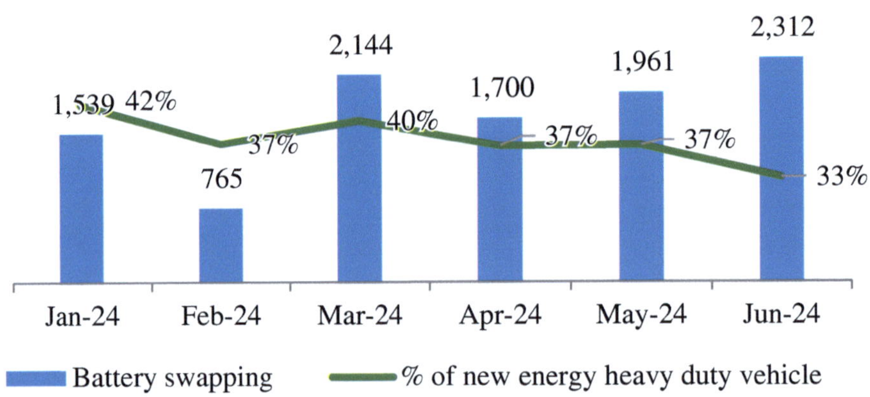

Figure 4.21 Sales volume and share of battery-swappable heavy trucks in new energy heavy trucks in 2024 (Units, %).

Entering 2024, the sales structure has experienced significant changes. As illustrated in Figure 4.21, while the monthly sales volume of battery-swappable heavy trucks demonstrated an overall upward trend throughout 2024, the market share of pure electric heavy trucks has seen a decline. If we consider the period starting from October 2023, when this decline first became evident, it is noteworthy that this downward trend persisted for ten consecutive months by June 2024. This situation warrants careful reflection on the underlying factors contributing to this phenomenon.

Tangshan is a prominent heavy industrial city in China, characterized by its numerous mines, steel plants, and ports. These factors create favorable conditions for the development of new energy heavy trucks. Since 2021, Tangshan has been one of the earliest pilot cities to initiate the transition to battery-swappable heavy trucks. The progress made in this area within Tangshan reflects the broader landscape of battery-swappable heavy truck development across China. Insights into future advancements in this sector can be derived from these developments.

According to research conducted by Che Shang Bang, the initial adopters of heavy truck replacements were large conglomerates—many of which also integrated logistics services and investments in new energy technologies—as well as leading players in battery swapping investments such as State Grid Investment and local governments that actively promote green transportation initiatives. The promotion of the dual carbon strategy was a key driver.[1]

These entities possess substantial financial resources and extensive operational capabilities, enabling them to establish their own battery swapping stations. Through collaboration among various stakeholders, they have achieved efficient battery swapping processes, cost-effective transportation solutions, and reliable cargo supply chains—thereby creating a closed operational loop.

To comply with environmental regulations and benefit from policy incentives, steel mills were among the first industries to mandate that 100% new energy heavy trucks be utilized for all entry and exit operations at their facilities; this was soon followed by similar requirements from ports and mining areas. Consequently, these sectors gained priority access on roadways for new energy heavy trucks. This phase marked the completion of the initial wave promoting battery-swappable heavy trucks—a period defined by explosive growth within this segment.

Entering 2023, the influence of small and medium-sized fleets and individual operators on the sales of new energy heavy trucks began to rise significantly, marking the onset of a second wave of promotion for the battery-swappable heavy truck market. Typically operating with fewer than 20 vehicles, these smaller entities collectively represented a substantial portion of the total heavy trucks in Tangshan.

However, they faced several challenges, including unstable cargo sources, low freight prices, inconsistent battery swapping standards, insufficient battery swapping infrastructure, diminished residual value of battery-swappable trucks due to limited ownership, high risks associated with batteries, and elevated costs related to battery leasing. Consequently, these operators often lacked confidence in adopting a lifecycle perspective. In their decision-making processes, short-term risks appeared to overshadow long-term benefits.

According to another survey by Che Shang Bang,[2] a fleet user stated, "If we lease batteries, we do not own them. Each vehicle would need to pay around 7,000 yuan per month for battery rent (for a battery product of about 220 kWh). While the initial cost for purchasing a battery-swappable heavy truck is lower upfront—around 300,000 yuan—the question remains: how much will we ultimately pay for renting the battery? For what duration? Acquiring a battery means it becomes our asset that can be sold alongside the truck later on. However, if we opt for battery swapping only to retain just the chassis without any accompanying batteries left

[1] CVZONE "Retalk the market development prospects of charging and replacing heavy trucks."
[2] CVZONE "Research: Why did the electric heavy truck fail to become mainstream in Tangshan Port."

behind—who would be interested in purchasing such a used truck?" Another fleet user said, "Compared to charging models, battery swapping is too costly. Including battery leasing fees, service fees, and electricity charges. But the minimum "starting fee" of several thousand yuan per month is just the baseline; for example, if we exceed 10,000 kWh of charging this month, every additional kWh will incur multiple times the cost (because frequent charging and discharging reduce the battery's lifespan, and the battery leasing company will multiply the cost after exceeding the "starting fee"). In this case, running 14,000–15,000 km per month could raise the battery rent from 7,000 yuan to 9,000 yuan. However, with a charging model, we can choose high-capacity products and charge during off-peak hours, which is economically more beneficial than battery-swappable heavy trucks."

From the feedback received, it is clear that risk aversion significantly influences the decision-making process. Currently, the Chinese government and relevant enterprises are collaborating to promote battery swapping stations, standardize battery swapping protocols, and enhance battery technologies. This concerted effort actively fosters the development of the heavy truck industry with swappable batteries.

In March 2024, Xu Changming, a senior economist at the National Information Center, highlighted through an analysis of freight volume data that "the development of heavy trucks is closely tied to real estate; approximately 40% of commercial vehicle sales are linked to this sector." Since 2021, indicators such as real estate investment, residential sales area, and new construction area have experienced negative growth trends that persist today. This trend is unlikely to change in the short term and will significantly impact the commercial vehicle market.

Based on this analysis, and in conjunction with the development of China's heavy truck industry, a forecast for the battery-swappable heavy truck sector has been formulated, as illustrated in Figure 4.22. By 2026, China is expected to complete the establishment of battery swapping facilities for trunk networks. Following the sales peaks observed in 2021 and 2022, a second wave of sales surge is anticipated during the period from 2027 to 2028.

By 2030, as autonomous driving technology matures and replacement demand sharply declines, the market for battery-swappable heavy trucks will likely experience a sales trough. Subsequently, with widespread adoption of autonomous driving leading to significant reductions in human labor requirements, overall operational efficiency for heavy trucks will improve while the number of active heavy trucks is projected to decrease annually.

China's advancements in battery-swappable models have occurred earlier than many competitors; these models feature relatively advanced technology and are witnessing an annual increase in exports. Ultimately, it is expected that traditional heavy trucks will be largely replaced by new energy variants at a rate of approximately 90%. Furthermore, a third wave of sales peak is anticipated around 2035.

Shanghai Qiyuan Green Power serves as the primary entity responsible for the construction of the Green Electricity Transportation Industry Innovation Center

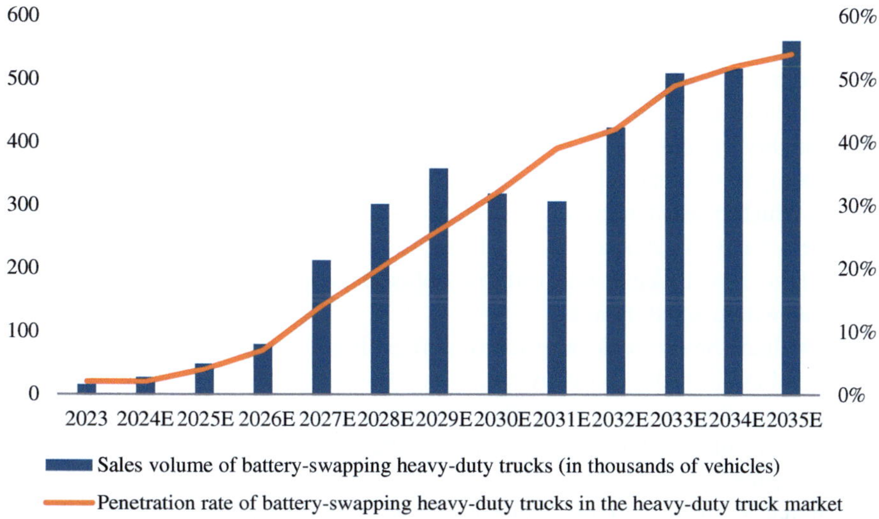

Figure 4.22 Forecast of sales volumes and penetration rate of battery swappable heavy trucks in China (thousand units, %).

under the auspices of State Power Investment Corporation. The company has independently developed a suite of operational management platform clusters, including the Vehicle Operation Management Cloud Platform. Additionally, it has created a diverse range of products such as onboard communication terminals, battery communication terminals, and mobile swap station control systems, which have been successfully promoted and implemented in over ten thousand units across more than 40 host vehicle manufacturers. In collaboration with 61 leading domestic heavy truck OEMs and three swap station manufacturers, Qiyuan Green Power has established a product line comprising 56 highly versatile electric models along with various fixed, mobile, and skid-mounted swap stations. By 2023, Qiyuan Green Power had successfully delivered over 23,000 battery-swappable heavy trucks and construction machinery into active operation. Notably, on average, five out of every ten battery-swappable heavy trucks available in the market utilized services provided by Qiyuan Green Power. Additionally, it constructed over 580 charging and swapping stations nationwide, a year-on-year increase of 132%, covering all provincial administrative regions of mainland China, with a market share exceeding 65%, making it the most mainstream heavy truck charging and swapping service provider in the country. In 2023, Qiyuan Green Power independently developed and produced its first shared battery system compatible with over 95% of the swap stations in the market, enabling cross-brand vehicle sharing and interchangeability, covering over 85% of battery-swappable heavy truck models.

Zhiliwulian Technology is an industrial enterprise dedicated to the commercialization of scientific achievements from Academician Minggao Ouyang's workstation at Tsinghua University, specifically in the realm of heavy truck battery swapping equipment and safe operation of battery big data. The company positions itself as a comprehensive integrator of digital battery swap operation solutions, providing both battery swap equipment and onboard systems that adhere to automotive interchangeability standards.

The company has launched China's first integrated side-swap heavy truck swap station, which boasts a remarkable swap time of just three minutes and accommodates between 4 and 20 spare battery bays. Additionally, Zhiliwulian has pioneered lightweight, high-locking, and durable onboard swap systems that comply with automotive industry standards.

Zhiliwulian's heavy truck swap stations officially entered the market in June 2023. By December 2023, the company had secured over 50 orders for its swap stations and successfully delivered more than 30 units. This achievement resulted in a market share of 10% for the year, which increased to an impressive 30% by the fourth quarter.

Shanghai Enneagon Energy Technology Co., Ltd. is a high-tech enterprise characterized by its "specialized, refined, unique, and innovative" approach to the research, development, production, sales, operations, and services of charging and swapping equipment for new energy electric vehicles. Enneagon is dedicated to pioneering electric vehicle charging and swapping products as well as comprehensive solutions. The company is committed to establishing an IoT-based ecosystem cloud platform for electric vehicle charging and swapping services.

By the end of 2023, Enneagon Energy had successfully deployed 402 heavy truck charging and swapping stations across 31 provinces and cities in China, covering a total of 97 cities nationwide. Additionally, Enneagon Energy has sold over 40,000 power battery packs that are compatible with more than 550 electric heavy trucks from over 40 different host vehicle manufacturers.

In 2022, the Module to Bracket (MTB) technology was introduced. The essence of this innovation lies in the direct integration of battery modules into the vehicle's frame or chassis, thereby optimizing the battery system and enhancing overall vehicle performance. MTB technology significantly improves spatial utilization within the battery system, achieving a volumetric efficiency increase of 40%. Additionally, by positioning the battery system lower within the vehicle structure, MTB technology reduces the center of gravity by approximately 21%, which in turn enhances handling stability and driving dynamics.

The implementation of a U-shaped water-cooling system further augments cooling effectiveness and cell efficiency, enabling high-power charging and discharging capabilities while supporting super-fast charging technologies.

Moreover, high-preload, no-impact locking technology ensures both safety and stability for the battery pack. In contrast to existing rear-mounted swap methods that utilize clearance fit without preload force—resulting in plastic deformation or metal fatigue over time—the new swap coupling technology employs a self-locking mechanism combined with a rubber pad to eliminate clearance issues. This advancement reduces total vibration energy by up to 35% and decreases vibration load on the battery pack by 50%, as illustrated in Figure 4.23.

The thermal management system technology employs an efficient heat dissipation design to minimize heat loss. Utilizing automotive-standard liquid-to-liquid quick-connect couplings, the water circuit of the battery pack exchanges heat with the vehicle's thermal management system, thereby achieving waste heat recovery—termed "integrated thermal management of vehicle and battery." Through the application of heat pump technology, waste heat from the power system is effectively harnessed, leading to a reduction in heating energy consumption for both the battery and occupants; this approach is particularly advantageous in extremely cold regions. Furthermore, as depicted in Figure 4.24, direct waste heat recovery constitutes 30% of total heating energy. This method features a straightforward structure and low cost, making it well-suited for northern regions.

In September 2023, a battery swap station located on Yihai Road in Tangshan City, Hebei Province, developed and equipped by Zhiliwulian, commenced operations. To accommodate the varying energy replenishment demands during the initial construction phase, Zhiliwulian proposed a phased construction plan, as illustrated in Figure 4.25.

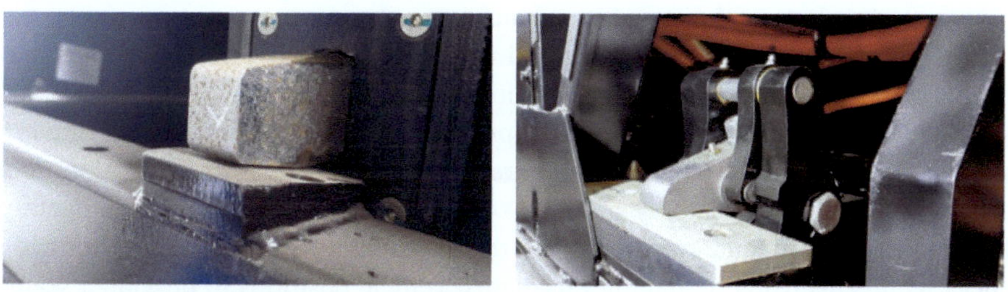

Figure 4.23 Existing market swap coupling solutions (left) high-preload coupling solution (right). *Source:* Zhiliwulian.

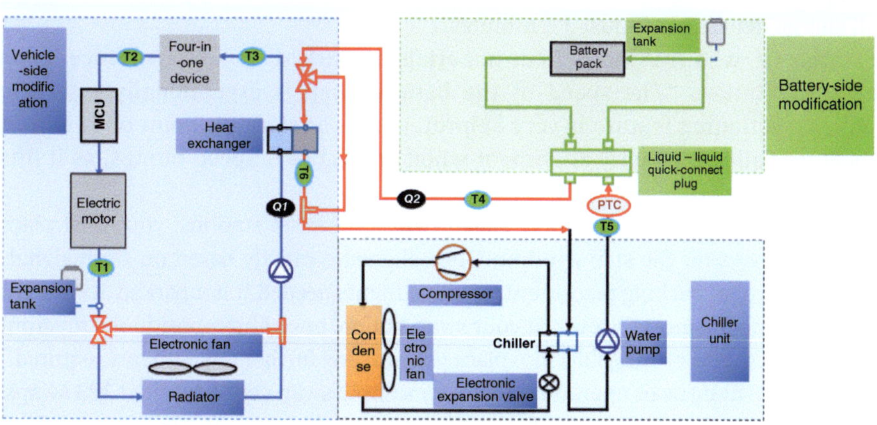

Figure 4.24 Direct waste heat recovery efficiency.

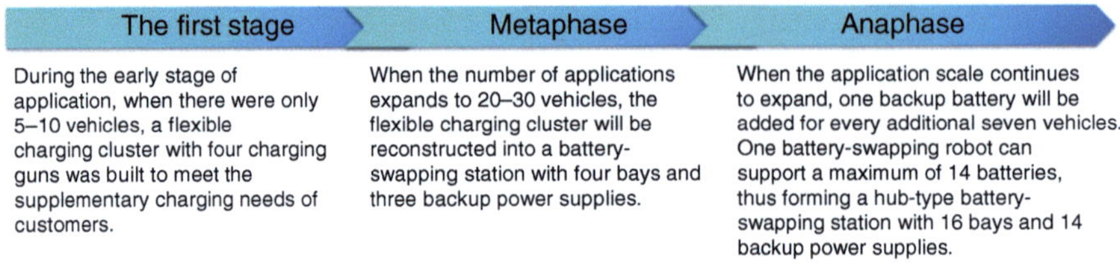

Figure 4.25 Construction expansion plan.

When the number of vehicles is limited—specifically with daily replenishment requirements below 10 vehicles—only charging guns are necessary for installation. As the demand increases to accommodate up to 30 vehicles requiring replenishment, an expansion can be made to include a battery swap station featuring four storage bays and three spare batteries. Furthermore, as vehicle numbers continue to rise, additional expansions can occur by adding one spare battery for every seven new vehicles introduced; each swap robot is designed to service up to fourteen batteries.

Based on preliminary research and calculations regarding the anticipated number of vehicles served, the final construction plan for this station encompasses seven spare batteries along with one intermediate storage bay.

The region is primarily served by semi-trailers transporting fiber materials and foreign minerals, characterized by high loads and intensive transport operations. After 14 days of operation, the station has processed a total of 887 orders, serving 176 vehicles with an average battery swap duration of 233 seconds per swap, alongside a cumulative electricity supply of 190,900 kWh. The swap station is intelligently designed to accommodate tractors, dump trucks, concrete mixers, and earthmoving trucks while effectively managing misaligned vehicle positions to provide a seamless swapping experience for drivers.

Drivers entering the station can effortlessly scan a code for automatic battery swapping; robots are equipped to intelligently identify the battery position, enabling full charge departures in just 3.5 minutes.

Master Qi, who transports fiber materials, remarked on his experience using the swap station: "The speed of the battery swap is exceptionally fast. The vehicle positioning feature is very helpful; there's a reference point on the front wheel that allows for easy alignment when driving over speed bumps, as it lifts slightly."

Master Ji from the Jiujiang fleet stated: "At other swap stations, you must place reference marks on the side window. Once aligned correctly based on horn signals indicating proper parking placement, or adjustments needed if not parked accurately, it can be cumbersome. However, at your swap station, once I have positioned my front wheel correctly, everything falls into place without any further adjustments required."

After three months of operation, the daily sample swap count reached 123 swaps, accommodating 96 vehicles, with an average swap duration of 226 seconds. Over a span of 98 days, the station successfully completed a cumulative total of 9,234 swaps, serving 198 vehicles. The average swap duration remained consistent at 225 seconds, with an overall average of 41 swaps per day.

4.3.2.1 Self-operated Swap Station

Revenue: Comes from battery swap service fees.

Costs: Include the operation and maintenance costs of the swap station.

Advantages: Full control over the operation of the swap station, direct interaction with customers, rapid response to market demand, improvement of service quality, and establishment of brand effects.

Challenges: Requires bearing higher operation and maintenance costs, involving aspects such as human resources, equipment maintenance, energy management, and land rent. It demands high management levels and technical support, and carries higher risks.

4.3.2.2 Joint Venture Model

Revenue: Shared investment with partners who have operational, technological, and policy advantages but do not participate in daily management.

Advantages: Stronger risk resistance, ability to leverage partner advantages to quickly establish a market presence.

Challenges: Difficulties in coordinating operations and management, need for clear profit distribution.

4.3.2.3 Entrusted Operation of Swap Station

Revenue: Comes from battery swap service fees.

Costs: Entrusted operation service fees of the swap station.

Advantages: Can outsource operational management to professional service providers, reducing management costs and pressure, allowing focus on core business, and facilitating quick project launch.

Challenges: Service quality and operational efficiency depend on the level of the outsourcing provider, which may lead to unstable service quality. Revenue from swap service fees needs to be shared with the service provider, potentially yielding less income compared to self-operation.

4.3.2.4 Leased Swap Station

Revenue: Fixed rental income.

Costs: Business management costs.

Advantages: Stable income and lower risk, no direct involvement in operation and management, reducing managerial complexity and operational costs, suitable for investors seeking stable returns.

Challenges: Rental income may be lower than that of self-operation or entrusted operation models, and may miss out on opportunities and higher potential returns brought by rapid market changes, unable to directly enhance customer experience and brand value.

4.3.2.5 Franchise Model

Revenue: Comes from battery swap service fees.

Introduction: A recently emerging business model for heavy truck battery swaps, where the franchisor provides brand authorization, uniform personnel training, and equipment installation, while the investor only needs to invest in basic equipment, site, and franchise fees.

Advantages: Lower initial investment, leveraging the franchisor's battery swap resources such as battery procurement, swap network, and customer base.

Disadvantages: Profits are subject to franchisor deductions, and the investor bears the business risk.

4.3.3 Strategic Recommendations for the Evolution of Heavy-duty Truck Battery Swapping

Currently, while battery-swappable vehicle models have achieved separate invoicing for the vehicle and the battery, announcements from the Ministry of Industry and Information Technology (MIIT) regarding these vehicle models still encompass both components. For vehicle registration purposes, it is necessary to present both the vehicle invoice and the corresponding battery invoice. This requirement significantly hinders the rapid promotion of a separation model for vehicles and batteries within the context of battery swapping. Consequently, this impedes not only the efficient circulation of swap batteries but also hampers the implementation of financial policies related to batteries.

It is anticipated that in future developments, an innovative announcement management model for vehicle-battery separation will be established based on existing MIIT announcements concerning battery-swappable vehicle models. This would involve creating a standardized system for interoperability interfaces related to battery swaps, substituting characteristics of battery systems with specifications pertaining to swap interfaces in official vehicle model announcements.

Furthermore, there should be an independent announcement management system dedicated specifically to managing announcements related to battery swap systems, as well as market access separately for vehicles. The use of compliant battery systems—sourced from various manufacturers with differing energy densities and capacities—that align with announced interface specifications should be permitted. This approach aims to achieve genuine separation between vehicles and batteries while facilitating efficient circulation of batteries as energy carriers.

Battery swapping for heavy trucks has emerged as the predominant technological approach to electrification, owing to its advantages such as vehicle-battery separation, rapid refueling times, minimal land use, and limited impact on the power grid. However, the implementation of the battery swap model encompasses numerous technical steps from design to execution, highlighting an urgent need for unification and standardization—particularly concerning battery manufacturing standards, interface specifications, and related software protocols for communication and identification between stations and vehicles.

The heavy truck battery swapping industry comprises a diverse array of stakeholders including equipment manufacturers, swap station operators, power grid providers, technology suppliers, battery manufacturers, energy storage facilities (battery banks), vehicle enterprises, end users (both companies and individuals), freight service providers, and financial service institutions. In the absence of authoritative leadership within this sector, establishing industry standards that equitably address the interests of all parties involved remains a significant challenge.

China's approach is spearheaded by state ministries and commissions, with industry standard leaders taking the initiative. Major enterprises are collaborating, pilot cities are implementing trials, and the influential roles of industry associations and alliances are being effectively utilized. This framework enhances demonstration projects focused on battery swap interoperability for electric heavy trucks as well as the development of interconnected swap station infrastructure.

Pilot cities aim to achieve standardization across various components including battery swap electrical interfaces, cooling interfaces, swapping mechanisms, battery packs, vehicle-battery communication protocols, station-vehicle communication systems, and charging processes at stations. To expedite the establishment of an interconnected swap network, it is advisable to encourage local governments to undertake initiatives aimed at formulating local standards for battery swap interoperability that align with automotive industry standards.

This strategy can address existing gaps in phased standards while providing validation data and practical experience that will contribute to long-term enhancements within the industry and national battery swap standards. Ultimately, this creates favorable conditions for advancing automotive industry standards toward national benchmarks.

The scale of battery assets for electrified heavy trucks is substantial, characterized by significant investment thresholds and advanced operational technology requirements. In 2023, cumulative sales of heavy trucks reached 910,000 units. If we project the future promotion of one million electric heavy trucks, an estimated 460 billion yuan worth of battery assets will be necessary, indicating a considerable asset scale.

Battery systems inherently pose safety risks; notably, the total electricity capacity of heavy truck swap batteries is fifty times greater than that of passenger vehicles. Given the complex and demanding operating conditions faced by heavy trucks and the diverse usage scenarios encountered by users, managing the safety of battery swap systems for these vehicles becomes increasingly intricate. Currently unresolved issues such as battery supervision, intelligent maintenance, and fire protection within heavy truck battery swap systems present challenges that could significantly impact the industry's development trajectory in light of any major safety incidents.

To address these concerns effectively, it is imperative to cultivate leading enterprises specializing in battery safety supervision and intelligent maintenance. Concurrently, local governments are encouraged to develop comprehensive monitoring platforms for battery safety to bolster oversight and facilitate intelligent maintenance throughout the entire lifecycle of batteries.

There is currently an absence of a revenue distribution mechanism and data exchange platform for swap service fees, battery rental fees, and electricity costs in the context of interoperable operations among different operators and battery banks. Consequently, even if swap stations comply with interoperability standards, they are unable to achieve seamless interoperability across various operators' stations and batteries from different battery banks.

It is recommended that leading enterprises be convened to negotiate and establish rules for the clearing and settlement processes related to interoperable operations for heavy truck battery swaps. Additionally, it is essential to collaboratively develop a clearing and settlement platform. Furthermore, it should be mandated that swap operators and battery banks offering public swap services must connect to this platform.

Under interoperable operational conditions, the challenge of determining responsibility for damage to swap interfaces is significantly heightened. Ambiguity regarding quality accountability can lead to delays in after-sales service for users and extended claim processing periods, ultimately resulting in unsatisfactory experiences.

To address these issues, it is recommended to establish a "Heavy Truck Interoperable Maintenance Emergency Service Fund" and develop a "Heavy Truck Interoperable Operation Quality Traceability Platform." This initiative would require swap operators and battery banks to connect with this platform, thereby ensuring enhanced after-sales service experiences within interoperable operational systems.

4.4 Ships: Technological Validation in Emerging Applications

4.4.1 Overview of Battery Swapping for Electric Ships

According to the Maritime Battery Forum's Ship Register (Maritime Battery Forum, 2024), there are currently approximately 1,300 battery-powered vessels operating globally. Figure 4.26 encompasses all types of ships, excluding privately used small leisure boats. Among these 1,300 vessels, 64% are classified as hybrid, 17% as plug-in hybrid, and 19% as fully battery-powered. Fully battery-powered and plug-in hybrid ships offer significant potential benefits in terms of emissions reduction. However, factors such as the nature of operations, operational areas, or insufficient charging infrastructure may hinder the regular charging of batteries using electricity from the grid.

The average size of battery systems installed on board ships has been steadily increasing. Over the past decade, the average battery capacity per vessel has risen

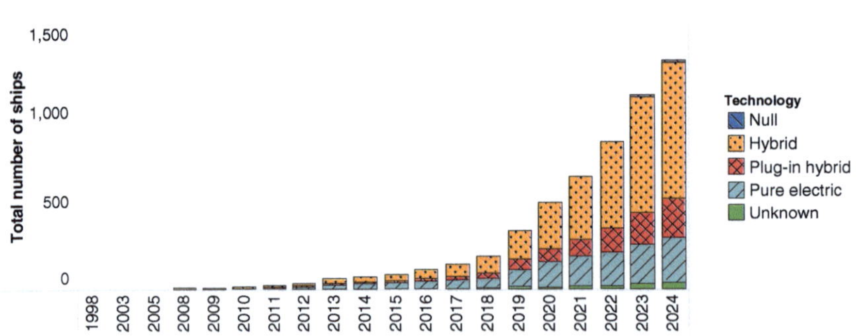

Figure 4.26 Uptake of maritime battery market. *Source:* Maritime Battery Forum.

from a range of 200–500 kWh to over 1300 kWh by 2024. The cost of maritime type-approved battery systems is approximately five times higher than that of automotive battery systems (Maritime Battery Forum, 2024). Consequently, the expenses associated with battery systems can constitute a significant portion of the overall costs involved in constructing a battery-powered ship.

With ongoing advancements and enhancements in maritime battery technologies, ship owners may find it challenging to determine the optimal timing for electrifying their vessels and selecting the appropriate battery technology for their needs. Additionally, the weight of these large battery systems can pose challenges on board ships, particularly for those operating at relatively higher speeds. This steady growth is illustrated by the annual data presented in Figure 4.27.

Battery swapping can therefore have several significant benefits for maritime applications compared to installing large, integrated battery systems.

Electric ships represent a category of electric transportation vessels. They primarily utilize battery storage systems or power generation units powered by renewable and clean energy sources, such as methanol, hydrogen, ammonia, and liquefied natural gas (LNG). Additionally, they may employ conventional fuel-powered generators as auxiliary power sources within an electrical system.

The electric ships referenced in this article are defined as vessels that are either wholly or partially powered by battery storage systems—commonly referred to as battery-powered vessels. These battery storage systems encompass lithium-ion batteries and energy-type supercapacitors.

Based on the vessel's cruising range and the configuration of the battery installation, electric ships can be further categorized into fixed-charging electric ships and battery-swappable electric ships. Fixed-charging electric ships have their battery storage systems installed in designated compartments onboard and are optimized for short-range voyages; these vessels require recharging at predetermined locations such as docks. In contrast, battery-swappable electric ships utilize container bodies

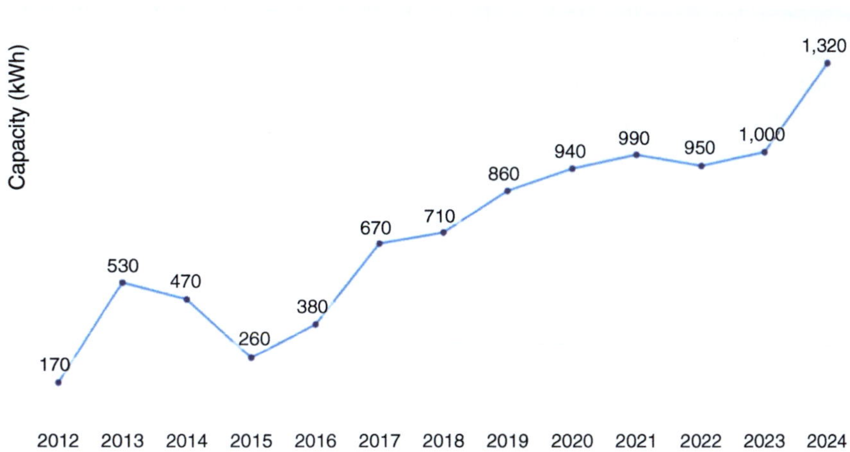

Figure 4.27 Average battery size per ship. *Source:* Maritime Battery Forum.

as platforms for mounting batteries. Their storage systems are integrated into standard containers strategically placed across various decks of the vessel. Designed for long-range travel, these vessels can recharge similarly to fixed-charging models at established locations but also allow for direct replacement of containerized battery packs to facilitate rapid power replenishment.

Both charging and battery swapping are essential methods for replenishing the energy of electric ships, each suited to specific scenarios. Charging is constrained by factors such as battery capacity, spatial arrangement on the vessel, and berthing time at the dock. Consequently, it is primarily suitable for short routes (not exceeding 300 km) and smaller tonnage vessels (up to 2,000 deadweight tons). Its contribution to promoting electric propulsion in transport vessels remains limited. In contrast, battery swapping offers several advantages: it reduces the initial cost of acquiring a vessel, alleviates range anxiety, enhances safety levels, and facilitates high-efficiency energy supply. As a result, it is anticipated that battery swapping will emerge as a critical method of energy provision for electric ships in the future.

It is important to clarify that containerized batteries typically refer to mobile power supplies housed within containers; these systems utilize container bodies as mounting platforms for their batteries. Battery swapping involves the rapid replacement of these containerized power sources using specialized equipment or manual assistance to replenish the electrical energy of battery-powered vessels. Currently, the development of containerized batteries follows a modular approach; this means that the power requirements of vessels can be met by combining modular containerized batteries, akin to building blocks.

The main application scenarios for electric ships include

Inland and coastal shipping: Electric Ships are suitable for inland transportation and coastal port operations, such as passenger ships, cargo vessels, and harbor working boats.

Sightseeing boats in scenic areas: In enclosed waters of scenic tourist areas, Electric Ships offer a quieter and more environmentally friendly sightseeing experience, as seen in locations like Wuhan's East Lake, Fujian's Cuiping Lake, Dali's Erhai Lake, and Xinjiang's Sayram Lake.

City ferries and sightseeing ships: Short-distance passenger transport between cities is another potential application scenario for Electric Ships due to their low noise and emissions, making them suitable for urban environments, such as night cruises along Fujian's Min River, the Two Dams and One Gorge area in Yichang, and night cruises in cities along the Yangtze River.

Government vessels: Ships used for patrol, search and rescue, and other official duties benefit from electric propulsion, which reduces noise and enhances stealth and mission efficiency.

The global electric vessel industry remains in its nascent stages, grappling with substantial initial investments, limited economic returns, and challenges such as range anxiety and insufficient shore-side power infrastructure.

First, there are technological bottlenecks that necessitate ongoing advancements in battery energy density, cruising range, and charging speed for electric ships.

Second, issues related to infrastructure development persist. Shore-based facilities for initiatives like the "Electrification of the Yangtze River" are still under construction, requiring considerable investment and time for both development and adaptation.

Lastly, policy support and market mechanisms require continuous adjustment and coordination. It is imperative that governments, enterprises, and research institutions enhance collaboration to collectively advance the development of electric ships.

Promoting electric ships not only contributes to environmental protection by reducing fuel consumption and greenhouse gas emissions, but also supports the achievement of significant goals such as the "Great Protection of the Yangtze River" initiative and the "Dual Carbon" targets. Furthermore, electric ships offer lower operational costs, enabling shipping companies to decrease their long-term expenses.

The electrification of vessels can also drive the development and innovation of related technologies, including battery technology, charging infrastructure, and electric propulsion systems. With the growing global demand for new energy solutions, the electric vessel market is anticipated to experience rapid growth, creating new opportunities for related industries.

4.4.1.1 Saving Operation and Maintenance Costs

Reduced fuel costs: Electric boats utilize electricity as their primary energy source, replacing traditional diesel or gasoline and significantly lowering energy consumption expenses. The unit cost of electricity is generally lower than that of fossil fuels, and electric boats exhibit higher energy conversion efficiency, further contributing to reduced operational costs.

Reduced maintenance costs: The design of electric boats features a simpler structure with fewer mechanical components in the engine. This results in decreased wear and tear as well as lower failure rates, thereby minimizing daily maintenance and repair expenditures.

4.4.1.2 Industrial Chain Driving Effect

Promoting related industries: The advancement and adoption of electric boats will stimulate the growth of related industries such as battery manufacturing, electric boat design and production, and charging infrastructure development. This industrial expansion not only generates numerous job opportunities but also enhances China's technological capabilities and market competitiveness within the realm of Electric Ships.

Promoting technological innovation: The research and development efforts associated with electric boats necessitate significant technological innovation. This demand will encourage universities, research institutions, and enterprises to increase their investments in R&D within the electric vessel sector, fostering technological progress and industrial upgrades. Such advancements can enhance the performance of electric boats while providing valuable insights for the electrification of other transportation modes.

4.4.1.3 Promotion of Related Industries

The electrification of vessels will provide a demonstration effect for the green transformation of the entire shipping industry. Shipping companies might draw on this experience to accelerate the electrification of their own vessels, enhancing the overall green development level of the industry.

4.4.1.4 Comprehensive Economic Effects

Initial investment and long-term returns: Although the initial investment in Electric Ships is high, their operational and maintenance costs are low. Moreover, as electric technology advances and battery costs decrease, the initial investment in electric boats will gradually reduce.

Social and economic benefits: The electrification of vessels presents a unique opportunity to not only reduce operational costs for governments and enterprises but also to yield substantial social and economic advantages. For instance, decreasing pollution emissions can lead to lower environmental remediation expenses, while the enhancement of ecological environments can significantly improve residents' quality of life. Furthermore, fostering the development of related industries has the potential to invigorate regional economies. The transition toward electric vessels will catalyze the green transformation within the shipping industry, thereby elevating the overall perception of sustainable shipping practices.

In summary, electric ships represent a vital strategy for phasing out fossil fuels and facilitating the dual circulation of domestic and international economies. They play an essential role in reducing carbon emissions and safeguarding environmental integrity while simultaneously promoting advancements in the new energy sector. This shift enhances both intrinsic dynamism and international competitiveness within our economy. In light of existing challenges, it is imperative that all stakeholders collaborate effectively to drive technological innovation and infrastructure development, strengthen policy support alongside market mechanisms, and ultimately achieve vessel electrification goals—contributing meaningfully to both China's green initiatives as well as global low-carbon development efforts.

4.4.1.5 Key Technologies for Battery Swapping in Electric Ships

The key technologies for battery swapping in vessels mainly consist of the following components:

1. **Battery System and Management Technology**

   Battery system: The battery system serves as the energy source for battery-swappable vessels, and its performance has a direct impact on the operational efficiency and economic viability of these vessels. Lithium batteries are currently the preferred choice for vessel applications due to their high energy density, extended lifespan, and superior safety characteristics.

   Lithium-ion batteries: These batteries exhibit high specific energy, specific power, and energy density, making them widely utilized in swap technologies.

   Lithium iron phosphate batteries ($LiFePO_4$): Compared to other lithium-based batteries, $LiFePO_4$ offers a longer cycle life and enhanced thermal stability.

Solid-state batteries: Representing a promising direction for future development, solid-state batteries theoretically provide higher safety levels and greater energy density.

Battery management system: This system monitors critical parameters such as voltage, current, and temperature of individual battery cells to ensure the safe operation of the entire battery pack.

Balancing control technology: This technology ensures that the state of charge is evenly distributed across the battery pack, thereby extending overall battery life.

Fault diagnosis and early warning: This feature guarantees the stability and reliability of the vessel's power system.

2. **Swap Station and Automation Technology**

Swap station layout: This section outlines the design of swap stations, taking into account various factors such as shipping routes, vessel density, and shore power supply. These stations can be strategically established at ports, docks, or riverbanks to facilitate easy access for vessels.

Automated swap technology: This technology employs automated equipment for the swift replacement of battery boxes, significantly minimizing docking and swapping times. Additionally, smart management through information technology enhances operational efficiency.

3. **Communication Technology Between Vessels and Swap Stations**

Data transmission technology: Wireless communication technologies, such as WiFi and Bluetooth, facilitate short-range data transmission, while mobile communication technologies, including 4G and 5G networks, enable long-range and stable communication.

Data exchange formats and protocols: The adoption of standardized data exchange formats and communication protocols enhances compatibility and efficiency in the interactions between vessels and swap stations.

4. **Safety Technology for Battery-swappable Vessels**

Integrated safety management systems: The integration of BMSs, vessel monitoring systems (VMSs), and other safety monitoring frameworks is essential to establish a comprehensive safety protection system.

Emergency response and disaster mitigation: It is crucial to develop detailed emergency response plans and risk control measures addressing potential incidents such as thermal runaway, short circuits, and other emergencies.

5. **Economic Analysis of Battery-swappable Vessels**

Battery cost: The primary determinant influencing the economic viability of battery-swappable vessels is the cost of batteries. Reducing manufacturing expenses while extending battery lifespan significantly enhances economic feasibility.

Operational costs: Effective operational management plays a vital role in minimizing maintenance and operational costs associated with swap stations.

6. **Arrangement of Battery Containers on Vessels**
   In accordance with the China Classification Society's "Guidelines for Inspection of Pure Battery-Powered Vessels," considerations regarding safety classification, arrangement, and installation of containerized power sources are of utmost importance. The design of battery-swappable vessels requires a comprehensive assessment of vessel characteristics, stability, structural integrity, fire zone division, safety passages, shore-based facilities, and the strategic placement of containerized power sources.
7. **Modular Containerized Battery Technology**
   Modular containerized batteries utilize lithium iron phosphate batteries as their energy carrier and are structured around a standard 20-foot shipping container as the assembly unit. These systems are characterized by high integration, large capacity, mobility, and versatility as modular power solutions. Each container typically comprises four functional compartments: the battery compartment, fire suppression compartment, power distribution compartment, and air conditioning compartment. Equipped with a lithium iron phosphate battery system along with ventilation and liquid cooling systems, fire suppression mechanisms, auxiliary electrical systems, external interface systems, and emergency cut-off features—these modules can be conveniently and safely connected or disconnected from the grid to facilitate rapid recharging for vessels.

   Leading battery manufacturers, including CATL, EVE Energy, and Gotion, have been actively engaged in the development and production of marine containerized batteries. For instance, CATL's 20-foot marine container battery boasts a capacity of approximately 1959.82 kWh. The container shell is constructed from two layers of steel plates filled with Class A fire-resistant rock wool.

   Internally, a 25 kW water-cooling unit continuously monitors and maintains the optimal temperature for the battery system. The nominal output voltage is XDC772V, capable of delivering six outputs at 250A DC each, thereby meeting stringent marine weatherproofing and watertight requirements. Additionally, it incorporates an active heptafluoropropane fire suppression system; each container has an approximate weight of 27 tons.

   Battery swapping is an innovative charging method utilized in the electric vessel industry, which facilitates the rapid detachment of containerized batteries from vessels. Conceptually akin to the detachable batteries found in certain mobile phones, this approach incorporates dedicated control interfaces alongside the power supply system. Currently, most electric ships position their containerized batteries at the stern.

   Under the battery swap model, users can easily transfer a depleted containerized battery from the vessel onto a container truck at the dock and replace it with a fully charged unit. This process enables swift recharging and significantly extends cruising range. Such a mode effectively addresses challenges associated with prolonged charging times and lengthy waiting periods for recharging, thereby enhancing operational efficiency and improving user convenience and experience.

However, it is important to note that current battery-swappable electric ships encounter several challenges. These include a multitude of diverse ship-container interface lines and nonuniform standards, which complicate battery swapping operations. Furthermore, incompatibility among different battery containers poses significant obstacles to the widespread adoption and application of battery-swappable electric vessels.

4.4.1.6 Current Development Status of China's Industrial Chain

1. **Battery**

   In the realm of aviation applications, the paramount characteristics of power battery assemblies are high safety and lightweight design. Furthermore, the takeoff and landing power output requirements for electric vertical take-off and landing (eVTOL) aircraft necessitate enhanced high-power performance. These attributes are manifested in key parameter indicators: the probability of thermal runaway and fire incidents within the battery system (indicating high safety), the energy density of the battery system (reflecting lightweight design), and the power density of the battery system (demonstrating high power capabilities). These indicators significantly differ from those associated with conventional power batteries, underscoring a need for newly developed aviation-specific power batteries.

   Given that large-scale mass production of eVTOL/UAM (urban air mobility) aircraft is still some time away, manufacturers of complete aircraft or batteries must incur research and development expenses upfront. As current outputs for eVTOL/UAM remain far below those seen in electric vehicles, it follows that unit prices for aviation power battery assemblies are considerably higher. Additionally, as projects involving complete eVTOL aircraft undergo rapid iterations, there is an imperative for supporting battery manufacturers to achieve agile development processes. In contrast to traditional suppliers who typically require over a year for product development cycles, some can achieve small-batch production in as little as six months—placing significant demands on suppliers.

   Moreover, battery manufacturers must adhere to stringent aviation standards and developmental protocols such as DO-311; this requirement further complicates customized development efforts while also increasing associated costs. Nevertheless, competition among domestic batteries tailored for eVTOL/UAM remains intense. Numerous battery manufacturers express willingness to engage in customized development initiatives specifically aimed at completing aircraft projects within this sector.

2. **Power System**

   Thanks to the advanced development of China's NEV industry, the motor, electronic control, and other subsystems for eVTOL/UAM in China possess inherent advantages and a solid foundation. Numerous domestic complete aircraft manufacturers are currently reliant on foreign suppliers for essential motors and drives to ensure the successful development of their aircraft products; for instance, they procure motors from companies such as Safran EngineUS and Magnix. In recent years, with the rapid growth and increasing maturity of this sector, an ever-growing number of motor and electronic control supporting enterprises have expressed their willingness to join the aviation supply chain.

3. **Avionics, On-board Equipment, and Airframe Structure Support**
   China's civil aircraft avionics equipment research and development foundation is relatively weak. There are very few companies with airworthiness engineering experience that employ the civil aircraft avionics development methods and processes recognized by FAA/EASA, such as ARP4754A. This situation is also linked to the historically limited number of commercial aircraft and traditional general aviation R&D projects in China.

   Regarding requirements, standards, and methodologies related to RTCA/EUROCAE—such as the guidelines for safety assessment processes of civil airborne systems and equipment outlined in ARP4761A; certification considerations for highly integrated or complex aircraft systems detailed in ARP4754B/ED-79; safety assessments for commercial aircraft specified in ARP5150/5151; development guidelines and certification considerations for Integrated Modular Avionics (IMA) systems found in DO-297/ED-124; design assurance guidelines for airborne electronic hardware presented in DO-254A/ED-80; software considerations for certifying airborne systems and equipment described in DO-178C/ED-12B; as well as environmental testing protocols for airborne equipment established by DO-160G—our industry generally lags behind our European and American counterparts in mastering these aspects.

   However, this capability can only be significantly enhanced through practical exercises involving model projects.

4. **Infrastructure**
   The infrastructure for eVTOL/UAM operations represents a significant area of interest related to complete aircraft systems. Within a short timeframe, numerous enterprises and capital institutions have emerged as key participants in this sector, primarily categorized as follows:

   1. Construction and operation units for takeoff and landing facilities/airports, predominantly comprising state-owned entities such as airport groups and government platform companies. Notable examples include the Shanghai East China UAV Base and Anhui General Aviation Holding Group.
   2. Planning, design, and construction of takeoff and landing field facilities. Prominent firms in this category include Zhejiang Shengxiang, Tianjin Hangda Zhi'an, and Guangdong Xinzheng.
   3. Research and development of air traffic control systems along with data management and digital infrastructure solutions. Key players in this domain are Tencent, the Guangdong-Hong Kong-Macao Greater Bay Area Digital Economy Research Institute, Zhuhai Anqing, and Kongzhong Tiaodong.
   4. Air traffic control service providers. In accordance with the national policy directives on low-altitude airspace management reform, comprehensive flight service general stations are established at the provincial level under stringent oversight. These stations may establish multiple Class A/B substations to deliver air traffic control services effectively. Representative

enterprises include Henan Air Traffic Control Industrial Development Co., Ltd., alongside the Zhejiang Low Altitude Flight Service Center.

4.4.1.7 Current Development Status of Overseas Industrial Chains

1. **Battery**

Power battery technology serves a crucial supporting role in the advancement of industries such as NEVs, maritime vessels, and robotics. Aviation power batteries represent a specialized category of power batteries, tailored to meet the specific requirements of aircraft and their operational environments. The development of these batteries is contingent upon the broader landscape of the large power battery industry. Globally, nations have prioritized advancements in power battery technology and established developmental objectives. The typical development trajectories are outlined as follows:

United States: In 2023, the United States Advanced Battery Consortium (USABC) revised its goals for advanced battery technologies with an emphasis on fast-charging capabilities. The objective is to achieve an 80% state-of-charge (SOC) charging time within 15 minutes while attaining a single-cell energy density of 275 Wh/kg and reducing cost targets to $75/kWh.

South Korea: According to the roadmap developed by the Korea Battery Industry Association in 2018, there has been a deceleration in the growth rate of battery energy density since 2020; however, significant enhancements have been observed in metrics such as battery lifespan. By 2025, it is anticipated that single-cell energy density will reach 330 Wh/kg with cycle life extending up to 1000 cycles. Traditional suppliers within the large aviation sector engage in research and development based on proprietary technologies while collaborating with relevant governmental bodies; Safran exemplifies this approach. Additionally, start-up companies often emerge from academic institutions or larger enterprises through incubation processes; Evolito serves as a notable example.

China: Guided by the "Technology Roadmap for Energy-saving and New Energy Vehicles" released in 2017, China has made significant advancements in power battery technology. In 2019, the energy density of single cells reached 240 Wh/kg (hard case) and 260 Wh/kg (soft package), while costs decreased to 1 yuan/Wh. It is anticipated that the target of achieving a single-cell energy density of 300 Wh/kg for power batteries will be met by 2020.

During the period of the 13th five-year Plan, China established a key R&D special project focused on NEVs (2016–2020). This initiative supports technological research and development for power batteries across six critical areas: new materials and systems for power batteries, high-energy-density lithium-ion batteries, high-power and long-life batteries, power battery systems, high-energy-density secondary batteries, as well as testing and evaluation methodologies.

The specific energy of industrialized lithium-ion batteries has surpassed 300 Wh/kg with costs falling below 0.8 yuan/Wh. Furthermore, the specific energy of next-generation lithium-ion batteries has exceeded 400 Wh/kg, while that of new system batteries has reached above 500 Wh/kg. According to the development goals outlined in "Made in China 2025," it is projected that the specific energy of single-cell batteries will achieve levels of 300 Wh/kg by 2020, increase to 400 Wh/kg by 2025, and reach an impressive target of over 500 Wh/kg (cell energy density) by 2030.

2. **Power System**

Most eVTOL aircraft power systems utilize the Distributed Electric Propulsion System (DEP), while a few consider alternative options such as hydrogen power or hybrid systems combining oil and electricity. Although electric motors offer advantages including high power density, simple structure, low maintenance requirements, and minimal spatial constraints—thereby mitigating many mechanical structural issues—the relatively low energy density of batteries results in a higher weight proportion for the power system compared to traditional aircraft. As motors have only recently emerged as viable power devices for airworthy aircraft, nearly all motor models currently lack airworthiness certification. Civil aviation authorities and industries across various countries are collaborating to expedite this certification process.

Given that electric drive represents an emerging form of propulsion technology, overseas enterprises providing such products can be categorized into several types:

1. Traditional large aviation industry suppliers engage in research and development as well as design based on their proprietary technologies and collaborations with relevant departments. A notable example is Safran.
2. Start-up companies often originate from universities or large corporations through incubation processes. A representative example is Evolito.
3. Large enterprises may acquire specialized companies or establish joint ventures to enhance their capabilities in this domain. An illustrative case is Rolls-Royce's acquisition of Siemens' electric and hybrid aviation propulsion business, eAircraft, positioning it as a supplier of electric and hybrid propulsion systems for aircraft.

3. **Avionics and On-board Equipment Support**

eVTOL aircraft prioritize high levels of automation, seamless coordination with air traffic control systems and operational management systems, as well as the comprehensive application of cutting-edge information technologies such as artificial intelligence and cloud computing. They also incorporate an increased number of sensors, enhanced overall health monitoring of the aircraft, and utilize both air-ground data links and satellite communication links. These technological advancements and innovative applications have created significant development opportunities for numerous start-up technology companies. Consequently, this sector exhibits several distinct characteristics:

Traditional onboard equipment manufacturers continue to play a crucial role; emerging start-ups are vibrant and actively seeking opportunities across various sub-segments; many large enterprises from other industries have entered this field to compete for this expanding market.

Traditional on-board equipment that supports enterprises possesses extensive experience in the research and development of airworthy systems. Most eVTOL aircraft manufacturers typically select their core avionics components to mitigate risks and expedite the aircraft certification process. These companies often maintain a comprehensive product portfolio, with notable examples including Honeywell and Garmin.

A prime illustration of an enterprise transitioning into this sector from other industries is Inmarsat, which primarily specializes in maritime satellite services and satellite communication systems. Given that the latest generation of aircraft requires satellite communication capabilities, Inmarsat has naturally ventured into the eVTOL support market.

Most emerging start-up firms concentrate on areas such as autonomous flight, artificial intelligence (AI), and network communication security, functioning as Tier 2 suppliers. For instance, Daedalean's AI-driven autonomous flight products have garnered clients like Honeywell within the avionics industry. Additionally, there are a few companies that directly cater to complete aircraft manufacturers; one such example is Near Earth Autonomy, which provides direct services to Volocopter.

4. **Infrastructure**

The successful implementation and application of eVTOL aircraft are inherently linked to the support provided by infrastructure. This primarily encompasses several key categories, including airports/takeoff and landing fields, charging stations, hydrogen refueling facilities, air traffic control systems, meteorological services, and digitalization efforts.

In the realm of airport facilities, there are predominantly two types of enterprises: (1) operational management units of traditional civil aviation airports and large heliports; and (2) technology innovation companies. The former typically controls passenger flow resources and may even hold a monopoly on specific geographical locations. Existing infrastructures can be adapted to accommodate eVTOL operations. These entities possess substantial capital reserves along with established operations and management teams that have a comprehensive understanding of civil aviation regulations, standards, and procedures. Generally speaking, they enter this market through collaborations with the latter type of enterprise. Notable examples include Frankfurt Airport Group in Germany, Munich Airport, and London Heathrow Airport in the United Kingdom.

Conversely, technology innovation companies view eVTOL takeoff and landing fields as significant opportunities for their development. They demonstrate a willingness to engage proactively with complete aircraft manufacturers as well as regulatory authorities to comprehend customer needs alongside regulatory policy requirements. Furthermore, they actively participate throughout the entire process, encompassing site selection, planning, design, construction, and operation of eVTOL takeoff and landing facilities.

The enterprises involved in eVTOL charging infrastructure primarily include power grid companies, electrical equipment manufacturers, and airport operators. Similar to the electric vehicle sector, there is currently no standardized charging protocol for eVTOL aircraft. For instance, Joby Aviation and Beta Technologies have established distinct standards. Consequently, various types of charging stations must be installed at takeoff and landing sites to accommodate their respective aircraft. The main international standards encompass the CCS (Combined Charging System) standard spearheaded by the General Aviation Manufacturers Association (GAMA), with key supporters including Archer Aviation, Beta Technologies, Volocopter, Overair, Wisk Aero, Eve Air Mobility, and Lilium; as well as the GEACS (Global Electric Aviation Charging System) standard developed by Joby Aviation.

In contrast to charging infrastructure development, progress within the global hydrogen energy industry chain has been comparatively slower, with even fewer supporting facilities available. The enterprises focused on eVTOL hydrogen refueling infrastructure are predominantly comprised of entities from the hydrogen energy industry chain along with other energy transformation companies.

The operation of eVTOL aircraft involves conducting normalized, high-traffic manned flight activities within low-altitude airspace above densely populated areas. This represents a significant departure from traditional civil aviation operations, necessitating distinct technical requirements. It demands more sophisticated and grid-based low-altitude weather forecasting support. Consequently, enterprises that have developed advanced algorithms, predictive models, and meteorological lidar technology are emerging.

Digital/IT infrastructure is another critical aspect of eVTOL aircraft. A defining characteristic of these vehicles is their high levels of automation, autonomy, and digitalization. To facilitate the development of application layer capabilities, substantial IT infrastructure support is essential—this includes cloud computing, digital twins, and AI-generated content (AIGC).

Central Enterprise Practices and Critical Technologies /Research on the Development of New Business Models for Battery Swapping

Under the new business model of battery swapping, two operational models have emerged: "Self-owned Batteries and Self-operated Swaps" and "Leased Batteries and Third-party Operated Swaps."

Self-owned Batteries and Self-operated Swaps refers to a scenario where the modular containerized batteries, swap stations, and charging stations are all owned by the shipowner. Entities that typically adopt this model are integrated port and shipping group companies. The primary advantage of this approach is that operational costs can be fully controlled, ensuring reliable operational plans. However, a significant drawback is the high initial investment required.

In contrast, leased batteries and third-party operated swaps involve modular containerized batteries, swap stations, and charging stations being owned by battery leasing companies. This model is often utilized by small to medium-sized shipping companies as well as individual operators. The key benefit here is a lower initial investment; the procurement cost for the motive battery does not burden the end-user operating the vessel, nor does it necessitate installing a diesel main engine for vessel owners. Once electric vessel standards are unified, initial construction costs for vessel operators can be substantially reduced. Nevertheless, this model presents challenges such as uncertain and potentially higher operational costs compared to conventional vessels, while making the effective guarantee of operational plans more difficult.

4.4.1.8 Development Trends of Electric Ship Battery Swapping Mode

The development trend of the electric vessel battery swapping model is intricately linked to the global shipping industry's transition toward sustainability. As environmental regulations become increasingly stringent and pressures regarding carbon emissions mount, traditional fuel-powered vessels are progressively being supplanted by more environmentally friendly Electric Ships. The battery swapping model for vessels represents an innovative operational approach for Electric Ships, characterized by rapid battery exchanges during voyages, thereby ensuring continuous and efficient navigation—an aspect that is particularly crucial for long-distance journeys.

4.4.1.9 Impact of Battery Swapping Vessels on the Vessel Market

1. Mitigating concerns over energy consumption and range: The advancement of battery swapping technology will significantly mitigate concerns related to energy consumption and range limitations for Electric Ships, gradually bridging the convenience gap between electric and diesel-powered vessels. In light of rising oil prices, heightened environmental awareness, and supportive national policies, the proportion of battery-swappable vessels is expected to increase steadily.
2. Usage and maintenance issues: Unlike conventional battery-charging vessels, those utilizing a battery-swapping system employ a centralized approach for storage, charging, and uniform distribution of power source components. Users are only responsible for covering corresponding electricity costs without having to contend with issues related to battery aging or maintenance associated with charging.
3. Impact on the traditional diesel-powered vessel market: The evolution of battery swapping technology facilitates convenient power replenishment options that could lead to an uptick in the number of electric-powered vessels while potentially decreasing reliance on diesel-powered counterparts. This shift may accelerate China's transition toward a low-carbon economy.

4.4.1.10 Impact of Battery Swapping Vessels on the New Energy Vessel Market

The implementation of battery swapping technology in the maritime industry can elevate the development level of the sector in three keyways:

1. Technological innovation: The advancement of battery swapping technology facilitates the separation of vessels from their power supply, enabling battery manufacturers to concentrate resources on enhancing the overall safety and energy efficiency of battery systems. Ship design and construction will only require integration with this power module, thereby accelerating the pace at which new energy vessels are upgraded.

2. Demand structure: By decoupling vessels from their power supply through a leasing model for batteries, initial investment costs for acquiring vessels can be significantly reduced. Furthermore, supported by various policy initiatives, battery-swappable vessels gain a competitive pricing advantage. Leasing batteries is more economical than outright purchase and allows for improved maintenance and upgrades, thus providing shipping companies with enhanced convenience associated with equipment advancements while bolstering the competitiveness of battery-swappable vessels.
3. Industry expansion: Green vessels represent a burgeoning industry supported by the state, exhibiting significant development potential. The primary advantage of advancing battery swapping technology lies in its ability to address safety and convenience issues associated with energy replenishment. The slow charging speeds of existing batteries, coupled with recent safety concerns, have posed substantial barriers to the widespread adoption of new energy vessels. Battery swapping technology facilitates a plug-and-play approach, effectively resolving challenges related to the convenience of energy replenishment. Timely battery exchanges and upgrades can mitigate safety risks stemming from battery aging. By addressing these critical safety and convenience issues, we can transform the factors that currently hinder the growth of green vessels, thereby increasing demand for battery-swappable ships. This demand-driven expansion has the potential to encourage companies to scale operations and create broader developmental opportunities for China's green shipping industry, underpinned by sustainable principles.

4.4.1.11 Study of Representative Cases

On December 8, 2022, the construction of a 700 TEU Yangtze River mainline electric container ship commenced in Yangzhou, Jiangsu. This vessel is classified as a 10,000-ton class ocean-river direct ship powered by containerized mobile power sources. It measures 119.8 m in length and 23.6 m in width, with a depth of 9 m and a designed draft of 5.5 m. The ship is equipped with 36 twenty-foot equivalent unit (TEU) containerized power sources and two AC asynchronous propulsion motors rated at 900 kilowatts each.

The designed cruising speed of the vessel is set at 10.5 knots per hour, offering a single-trip range of approximately 380 km. With the integration of shore-based battery swap technology, this innovative design facilitates segmented range extension through the establishment of swap stations at key ports such as Wuhan and Nanjing; each battery swap operation takes between 20 and 30 minutes.

This project marks the first instance in China where a fleet operates a 10,000-ton class ocean-river direct vessel powered by containerized mobile power sources and supported by shore-based battery swaps along the Yangtze River mainline. The ship represents significant advancements in loading capacity and operational range, signifying an important transition from research and development to practical application within the realm of inland electric vessel battery swapping technology. A photograph of this innovative vessel is shown in Figure 4.28.

Figure 4.28 700 TEU-class Yangtze River mainline electric container ship.
Source: WeChat Official Account.

In February 2022, construction commenced on Hubei's inaugural 120 TEU new energy pure electric container demonstration vessel, the Hanjiang 120 TEU pure electric bulk/container dual-purpose ship. The Hanjiang 120 TEU is equipped with dual propulsion motors powered by batteries, measures approximately 80 m in length, and primarily operates along the Yangtze River and the Han River. Its maximum capacity is 120 TEU of 20-foot containers, featuring four containerized power sources with a total output of 1935 kWh each installed at the stern, amounting to an overall battery capacity of 7740 kWh. The vessel boasts a single-trip range of up to 175 km and can sustain operations for five days without recharging. It embodies "green, low-carbon, and environmentally friendly" characteristics that effectively replace traditional fuel consumption while significantly reducing harmful gas emissions. This innovation contributes to the standardization of shipping routes and advances new models within the container transport industry, thereby promoting sustainable development. The photograph of this new energy pure electric container demonstration vessel is shown in Figure 4.29.

Hubei's scientific and technological capabilities were harnessed in developing this vessel through the integration of various specialized technologies such as containerized energy storage stations, distributed DC power systems, and low-speed motor direct propulsion devices.

In November 2023, the first voyage of this battery-powered container liner operating on the Yangtze feeder route was successfully completed. When fully charged, its batteries store a total of 1,935 kWh of electricity; three fully charged mobile power sources can support a round trip along the Wuhan-Xiantao route covering a distance of 150 km. The separation between water and electricity mitigates safety hazards during shore charging operations; notably, the entire battery swap process requires only 10 minutes to complete. Based on an operational schedule

Figure 4.29 The Hanjiang 120 TEU. *Source:* WeChat Official Account.

involving approximately 84 round trips per year, the annual power consumption is estimated at around 530,000 kWh—equivalent to replacing 132 tons of fuel—and resulting in reductions in various gas emissions by 392 tons annually while achieving energy cost savings totaling 730,000 yuan.

In August 2022, China launched its first pure electric inland container ship, the "Jiangyuan Baishou," with a capacity of 120 TEUs. This vessel measures 79.92 m in length and 12.66 m in width, enabling it to transport up to 120 standard containers at a cruising speed of 17 km per hour. It operates entirely on battery power, utilizing lithium iron phosphate batteries as its energy source. The ship is equipped with two containerized power units located at the stern, each containing 1,540 kWh of lithium iron phosphate batteries. Additionally, it features a plug-and-play battery swap model for enhanced efficiency.

Compared to traditional vessels, this electric ship is projected to eliminate approximately 500 tons of fuel consumption annually and reduce harmful gas emissions by an estimated 1,520 tons. Jiangsu Province, recognized as a major waterway hub, boasts the Yangtze River extending east-west for 370 km and the Beijing-Hangzhou Grand Canal running north–south for 687 km; there are around 30,000 transport vessels operating within the province. With the promotion of pure electric ships, significant reductions in electricity costs for maritime operations are anticipated alongside economic benefits stemming from economies of scale.

In October 2022, following collaborative efforts between the Jiangsu Provincial Department of Transport and Jiangsu Port Group, the "Jiangyuan Baishou" commenced operations on the Beijing-Hangzhou Grand Canal as a pioneering example of battery-swapping technology for electric container ships. This milestone signifies the advent of an "electric era" in inland transportation within Jiangsu Province. This pioneering battery-swapping electric container ship is pictured in Figure 4.30.

In July 2022, China inaugurated its first "zero-carbon demonstration route"—a customized container route from Hefei to Wuhu. This marked the commencement of the Anhui Wuhu-Hefei 105 TEU electric container ship (battery swap) demonstration project. The vessel, named "Port Shipping Journey 01," measures 73 m in

Figure 4.30 Jiangyuan Baishou. *Source:* WeChat Official Account.

length, 13.8 m in width, and has a depth of 4.6 m, with a container capacity of 105 TEU and a designed cruising speed of approximately 12 km per hour.

Equipped with two containerized batteries, the ship utilizes battery swapping facilities established by Qiyuan Green Power at the ports of Wuhu and Hefei. This system is sufficient for completing a single round trip covering a distance of up to 290 km. As a fully electric vessel powered by lithium iron phosphate batteries, known for their high safety standards, it emits zero pollutants during navigation. This initiative significantly contributes to energy conservation, environmental protection, and ecological preservation efforts in Anhui Province. A photograph of the "Port Shipping Journey 01" is shown in Figure 4.31.

"Yuetong Zhujiang 001" is the inaugural oil-to-electric conversion cargo ship and the first electric container vessel in the Guangdong-Hong Kong-Macao Greater Bay Area. The ship measures 49.99 m in length, 13.00 m in width, and 4.40 m in depth, with a cruising speed of 13.2 km per hour and a maximum carrying capacity of 1,740 tons. It is equipped with two containerized power sources, each boasting a capacity of 1,935 kWh, resulting in a total lithium battery capacity of 3,870 kWh for the vessel. This innovative craft supports both charging and battery swapping operations and can navigate approximately 100 nautical miles on a single charge, establishing it as a truly "zero-emission, zero-pollution, low-noise" clean vessel. A photograph of the "Yuetong Zhujiang 001" is shown in Figure 4.32.

Upon its official operation commencement, it is anticipated that this ship will save around 71.28 tons of diesel annually while reducing CO_2 emissions by approximately 220.97 tons. As the first container ship within the Greater Bay Area to achieve green navigation through oil-to-electric conversion technology, this vessel serves as a replicable and scalable model for energy-saving conversions of similar diesel-powered container ships.

Figure 4.31 Port Shipping Journey 01. *Source:* WeChat Official Account.

Figure 4.32 Yuetong Zhujiang 001. *Source:* WeChat Official Account.

Utilizing EVE Energy's LF280K cell-integrated batteries—accredited with CCS type approval certificates and TÜV Rheinland explosion-proof certification—these cells exhibit an impressive cycle life exceeding 8,000 cycles; thus supporting marine battery systems with lifecycles extending beyond ten years. With such high-quality cell products at their disposal, EVE Energy can tailor battery solutions featuring varying capacities and voltages according to customer specifications and operational conditions while fulfilling clients' requirements for green practices that are low-carbon and environmentally friendly.

The newly constructed 208TEU electric ship is a 4,200-ton cargo vessel with an overall length of 88 m. Through optimized design and advanced integrated power systems, this vessel incorporates two core technologies: green propulsion and intelligent systems. This type of ship can significantly enhance loading capacity and cargo handling capabilities. With the establishment of a network of charging and swapping stations for ships, it is projected to achieve over 10% in comprehensive energy savings for shipowners. Once completed, it will stand as the largest pure electric dual-purpose ship in China regarding container capacity that can pass through standard locks at a length of 88 m, as well as being the first new energy pure electric dual-purpose vessel in Hunan Province. In the future, the company aims to collaborate with additional partners to harness synergistic effects within its ecosystem, thereby empowering the value chain of the green water transport industry and striving to establish a "Hunan model" for electric vessels.

Construction of China's first 208TEU pure electric dual-purpose ship, named "Electric Xiangjiang," commenced on June 12th in Xiangtan City through a joint initiative by Qiyuan Green Power and Xiangtan Electric Group. Upon completion, this vessel will be recognized as both the largest pure electric dual-purpose ship capable of passing through standard locks at an overall length of 88 m in terms of container capacity across China and also as Hunan Province's inaugural new energy pure electric dual-purpose vessel. A photograph of the "Electric Xiangjiang" is shown in Figure 4.33.

This innovative 208TEU electric ship not only facilitates battery swapping more efficiently than traditional refueling methods but also integrates digital technology alongside intelligent systems while reducing crew requirements. Following the development of a network comprising charging and swapping stations for ships, it is anticipated that this vessel will achieve over 10% in comprehensive energy savings.

Upon completion, the electric ship will operate on a 160-km route between Yueyang Chenglingji Port and Changsha Jixing Port in Hunan Province. It is anticipated to reduce CO_2 emissions by 900 tons annually, which is equivalent to the total CO_2 emissions produced by 330 passenger cars. Currently, there are approximately 10,000 vessels operating in Hunan Province, the majority of which rely on diesel engines for propulsion. To achieve the "dual carbon" goals, electrification of transportation represents a critical area of focus, with the development of electric ships being an essential component of this initiative.

Xiangtan, known as a cradle of red culture, is part of the Xiangjiang water system that connects Hunan Province's "one lake and four rivers," all flowing into the Yangtze River. This geographical advantage provides unique benefits for the shipping industry. As a comprehensive service provider for zero-carbon green energy transportation solutions, Qiyuan Green Power has innovatively adopted a business model characterized by "separation of ship and power," thereby developing an integrated "land-water electric ship charging and swapping solution."

In July 2022, Qiyuan Green Power successfully launched the battery-swappable electric container ship "Port Shipping Journey 01," developed in collaboration with Anhui Port Group. In October of that same year, they introduced "Jiangyuan Baishou," China's first pure electric container ship with a capacity of 120 TEUs—developed jointly with Jiangsu Ocean Shipping—which made its maiden voyage shortly thereafter. Through extensive market applications within medium-short range shipping segments characterized by high frequency and significant pollution levels, Qiyuan Green Power aims to accelerate the green transformation and

Figure 4.33 Electric Xiangjiang. *Source:* WeChat Official Account.

development of water transport while promoting national efforts toward achieving dual carbon objectives. This endeavor will provide robust support for establishing a green, environmentally-friendly, high-quality, and efficient water transport industry in Hunan Province.

The successful launch of the "Electric Xiangjiang" electric ship project represents a significant milestone in the electrification of inland shipping vessels within Hunan Province.

Looking ahead, Qiyuan Green Power will collaborate with Xiangtan Electric Group, Xiangshui Group, and other partners to harness the synergistic effects of ecosystem collaborators. This collaboration aims to enhance the application coverage of green transportation products, empower the value chain of the green water transport industry, promote high-quality development within China's water transport sector, and collectively establish a benchmark for a sustainable industry!

4.4.2 Overview of Battery Swapping in Electric Aviation

With the growth of the global population and the acceleration of urbanization, issues such as road congestion and air pollution in existing transportation systems have become increasingly severe. In this context, eVTOL aircraft have emerged as a promising solution.

The eVTOL technology utilizes electricity as its power source, thereby eliminating the pollution associated with traditional helicopters that rely on internal combustion engines. This advancement is significant for reducing carbon emissions and achieving sustainable development goals. Furthermore, the convenience offered by vertical take-off and landing capabilities presents new possibilities for future transportation solutions.

4.4.2.1 Current Development Status of eVTOL in China

The development of China's eVTOL and UAM complete aircraft has been significantly influenced by the release of Uber's white paper titled "Fast Track to On-Demand Urban Air Mobility" in November 2016. This document sparked a surge of investment in the capital sector and initiated extensive discussions within the industrial community. In 2017, several landmark events unfolded in China:

1. On November 13, Geely Automobile Group acquired Terrafugia, a US-based company, thereby officially entering the flying car industry. This acquisition ultimately led to a series of strategic initiatives by Geely in both the low-altitude economy and eVTOL sectors. These initiatives included investments in Volocopter, the establishment of a joint venture named Volant Aerotech, and the launch of its self-developed complete aircraft project AE200. This move underscores the significant attention and substantial investment from traditional Chinese industries as they seek to penetrate the eVTOL market while also highlighting China's international ambitions within this domain.
2. The research and development team at Autoflight was established in Shanghai, marking its entry into manned eVTOL research. In 2021, Autoflight successfully completed a financing round that raised $100 million. This achievement is noteworthy as it signifies that Chinese technology startups are making strides into the eVTOL field. Both Geely and Autoflight have subsequently had direct influences on enterprises such as TCab Tech and Volant Aerotech, along with attracting interest from various capital institutions.

More than 30 enterprises in China have ventured into the research and development of eVTOL (electric Vertical Take-Off and Landing) complete aircraft. According to incomplete statistics, currently, 15 manned eVTOL R&D companies have successfully secured financing exceeding tens of millions of yuan; another 15 have completed the production of 1:1 prototypes and are actively engaged in testing. The majority of China's eVTOL complete aircraft enterprises are technology-driven start-ups. Notably, five companies that were established less than three years ago have already progressed to the prototype testing stage.

In comparison with leading firms in Europe and the United States, Chinese eVTOL complete aircraft enterprises tend to be smaller in scale, possess a shorter operational history, and generally attract lower levels of financing. Consequently, they face a greater need for support from both government entities and capital markets, as well as for collaboration and mutual assistance within the industry as a whole.

4.4.2.2 Current Development Status of Overseas eVTOL

According to statistics from the vertical flight society (VFS) in the United States, by the end of March 2024, there were a total of 983 eVTOL concept products worldwide. Among these, 48% were multi-rotor types, 16% were lift + cruise types, and 36% were vector propulsion types.

Due to the incorporation of numerous innovative design features, eVTOL aircraft currently lack extensive engineering experience and reference cases. Consequently, many fundamental research and verification tasks must be undertaken by design teams—this process significantly differs from that associated with conventional general aviation aircraft. For a typical Part 23 general aviation aircraft, it generally requires between 200,000 and 600,000 man-hours to complete all work from project initiation through certification. In contrast, developing an eVTOL may necessitate hundreds of thousands or even over two million man-hours; this substantial demand also contributes to increased R&D costs. Many start-up companies have not fully recognized this reality, leading to repeated delays in achieving their established enter-into-service (EIS) goals.

Statistics indicate that only 29% of projects reach the scale model verification stage while merely 5% advance to full-scale prototype R&D stages. The period from 2024 to 2026 is poised to be critical for transitioning eVTOLs from research and development into large-scale commercial operations. Major industry leaders are planning to deploy their models during this timeframe.

4.4.3 Battery Swapping Trends in Electric Aviation and Marine Sectors

The electric aviation industry has experienced significant growth in recent years, propelled by the dual forces of environmental sustainability and technological advancement. Among various operational models, the battery swapping mechanism for electric aircraft has emerged as a pivotal area of interest, possessing the potential to transform the future of aviation.

4.4.3.1 Current Landscape of Electric Aviation Battery Swapping

Currently, several pioneering companies have embarked on projects focused on electric aviation battery swapping. These organizations are working diligently to address the technical challenges associated with swiftly and safely replacing depleted batteries during flight or between flights. For example, some experimental eVTOL models have incorporated preliminary battery swapping systems. Although these initial efforts remain in the testing and refinement phase, they have established a foundation for further progress.

4.4.3.2 Technological Advancements Propelling the Mode Forward

4.4.3.2.1 Battery Standardization

In terms of infrastructure, a limited number of battery swapping stations have been constructed, primarily located near test flight areas or key research facilities. These stations are equipped with robotic arms and advanced sensor technologies designed to ensure precise extraction and insertion of batteries, thereby minimizing human error while maximizing operational efficiency.

4.4.3.2.2 *High-energy-density Batteries*

The ongoing advancement in battery chemistry is driving the development of high-energy-density batteries. These sophisticated power sources are capable of storing greater amounts of energy within a specified volume and weight, which directly translates to extended flight ranges for electric aircraft. Lithium-sulfur and solid-state batteries are among the leading candidates in this research area. Once they achieve commercial viability, these technologies will transform the battery swapping paradigm by decreasing the frequency of swaps required, thereby enhancing overall operational efficiency in flight.

4.4.3.2.3 *Automated and Intelligent Swapping Systems*

The incorporation of artificial intelligence and automation technologies represents another significant breakthrough. Automated battery swapping systems can autonomously assess battery health, predict potential failures, and optimize the swapping sequence based on real-time flight schedules and energy demands of the aircraft. Robotic arms equipped with machine learning algorithms can adapt to various battery designs and environmental conditions, ensuring seamless and error-free operations even under complex scenarios.

4.4.3.3 Infrastructure Expansion and Integration

4.4.3.3.1 *Scaling Up of Battery Swapping Stations*

As the demand for electric aviation continues to rise, there is an urgent necessity to expand the network of battery swapping stations. These facilities will no longer be limited to specialized aviation zones; instead, they will gradually permeate urban centers and regional airports. Strategic partnerships are being established among electric aircraft manufacturers, energy companies, and infrastructure developers to finance and construct these stations at an accelerated pace. This expansion will provide essential support for commercial electric air taxi services and regional electric flights.

4.4.3.3.2 *Integration with Renewable Energy Sources*

To enhance the environmental sustainability of the battery swapping model, there is a growing emphasis on integrating battery swapping stations with renewable energy sources. Solar panels and wind turbines can be installed on or near these stations to generate clean electricity for battery charging. This synergy not only reduces the carbon footprint associated with electric aviation but also mitigates the impact of power grid fluctuations on battery charging operations, thereby ensuring a stable and sustainable energy supply.

4.4.3.4 Regulatory and Safety Enhancements

The regulatory framework governing electric aviation battery swapping is evolving in parallel with technological advancements. Aviation authorities worldwide are developing new standards and regulations aimed at ensuring the safety and reliability of battery swapping operations. These include stringent guidelines regarding battery handling, storage, transportation, as well as protocols for emergency response in case of incidents related to batteries. Additionally, safety enhancements are being implemented at a technological level; advanced fire suppression systems are being integrated into battery swapping stations to prevent and manage potential fires involving batteries. Furthermore, BMSs are being reinforced with redundant safety circuits designed to monitor and regulate performance while safeguarding against overcharging, overheating, and short circuits.

The development trends of the electric aviation battery swapping model indicate a future characterized by enhanced efficiency, safety, and sustainability. With ongoing technological advancements, infrastructure expansion, and robust regulatory support, this model is set to emerge as a mainstream solution in the realm of electric aviation. It promises not only to unlock the potential of electric aircraft for commercial applications but also to make significant contributions toward reducing the environmental impact of the aviation industry, ushering in a new era of green and intelligent air travel. As stakeholders across various sectors continue to collaborate and innovate, there are no limits to what electric aviation battery swapping can achieve.

4.4.3.5 Development Trends of Battery Swapping Ships

1. **Selection of swapping stations:** The placement of battery swapping stations is strategically determined to ensure that battery swappable vessels can efficiently access power supply within their operational range. The selection process must take into account various factors, including the vessel's mileage, docking conditions, and power distribution requirements. Swapping stations are tasked with supplying energy to a significant number of batteries; thus, integrating advanced renewable energy technologies can substantially lower the operating costs of these stations, alleviate strain on the electrical grid, and utilize clean energy sources for charging batteries—thereby enhancing overall energy efficiency.
2. **Birth of service networks:** The architecture of service networks primarily encompasses IoT frameworks, billing systems, communication positioning networks, and scheduling mechanisms for battery swaps. While swapping stations can connect directly to the local power grid—eliminating the need for fuel transportation—they still necessitate efficient logistics for battery delivery, emergency support services, and inter-station battery exchanges. For instance, if numerous vessels require battery swaps at a single station in a short timeframe, resulting in an inadequate supply of fully charged batteries due to prolonged charging cycles, it becomes essential to source additional charged batteries from nearby stations with surplus capacity. This scenario underscores the importance of an IoT distribution network linking multiple swapping stations while also considering their geographical locations. In the realm of battery-swappable vessel technology, it is crucial to establish connections among vessels, users, and swapping stations for seamless interaction regarding batteries as well as billing processes. Typically integrated with communication networks or other relevant service infrastructures, this billing system forms part of a comprehensive user service network. Users are empowered to engage with swapping stations through

mobile devices for functions such as billing inquiries, battery reservations, prepayments for services rendered, or settlements related to transactions—all aimed at facilitating smooth operations within this innovative ecosystem.

3. **Design of battery swappable vessels:** In addition to fulfilling the requirements of traditional power vessels, battery swappable vessels must pay particular attention to structural reinforcement beneath box-type power sources, hazardous areas, safety passages, vessel endurance, and battery power considerations.
4. **Combination of charging and swapping:** The combination of charging and swapping primarily encompasses swapping stations that integrate both charging and swapping services, as well as simple swapping stations that exclusively offer battery swapping without any charging facilities. In simple swapping stations, batteries are delivered in conjunction with nearby swapping locations; conversely, combined charging and swapping stations provide users with the flexibility to select either service based on their needs.

In summary, national policies serve as barometers for industry development and play a crucial role in shaping the evolution of battery-swapping models. Although battery-swapping technology has emerged later than charging pile technology, it possesses significant advantages in terms of convenience, safety, and consumer costs. This technology can effectively address current bottlenecks hindering the advancement of electric ship charging solutions while fostering green vessel technologies and stimulating growth within a new battery industry. As public demand increases alongside heightened environmental awareness supported by national policy initiatives, pilot projects for battery-swappable vessel docks are gradually being implemented. Battery-swappable vessels contribute positively to promoting low-carbon practices within the maritime sector due to their high feasibility and market potential. In the short term, there is promising market potential for these vessels in short-distance transportation across inland rivers and lakes; thus, the future appears optimistic.

4.4.3.6 Challenges Faced by Battery Swappable Ships

The primary challenges currently confronting battery-swappable vessels include:

1. **Breakthroughs in battery technology:** The dimensions, weight, and range of batteries are critical factors that hinder the development of battery swappable vessels. There is a pressing need for more efficient, lighter, and longer-lasting battery technologies.
2. **Infrastructure construction:** The establishment and maintenance of battery swapping stations necessitate significant financial investment and must ensure reliable operation under diverse weather conditions and at sea.
3. **Safety and regulations:** Ensuring the safety performance of batteries as well as operational safety during the swapping process requires thorough testing and certification. Additionally, relevant international maritime laws and standards are still undergoing refinement.

4. **Economics:** The initial capital expenditure for battery swappable vessels is substantial; therefore, it is essential to strictly manage operational and maintenance costs to enhance economic competitiveness.
5. **Environmental impact:** Although battery swappable vessels contribute to reducing greenhouse gas emissions during operation, the production and recycling processes associated with batteries may also have environmental implications, highlighting the necessity for sustainable production methods and recycling technologies.

Overall, the development prospects for battery-swappable vessels are promising; however, significant efforts must be directed toward technological innovation, infrastructure development, regulatory frameworks, and economic considerations to address existing challenges.

4.4.3.7 Key Technologies for Battery Swapping in Electric Ships and Core Enterprise Practices

The market participants engaged in electric vessel battery swapping encompass vessel operators, battery swap service providers, and energy suppliers, among others. Their service model is illustrated in Figure 4.34. These entities interact through information exchange, ultimately establishing a multi-party coordinated business model for electric vessel battery swapping. The sustainable operation of this business model relies fundamentally on the reasonableness of the battery swap solution.

4.4.3.8 Practices of Core Enterprises

4.4.3.8.1 China Energy Engineering Corporation Storage Technology Co., Ltd. (Wuhan)

China Energy Engineering Corporation Storage Technology Co., Ltd. (hereinafter referred to as Zhongchu Technology) is a specialized provider of energy storage and battery system solutions, as well as technical services, under the auspices of

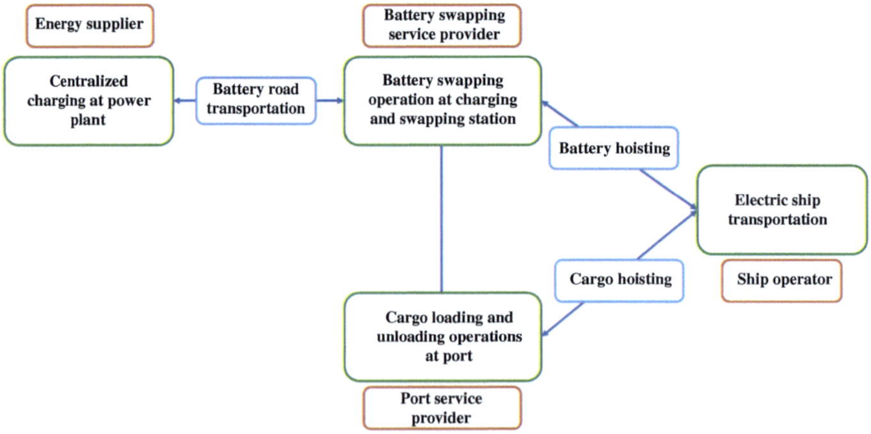

Figure 4.34 Schematic diagram of the electric ship battery swapping service mode.

China Energy Engineering Corporation. In alignment with the "30·60" dual carbon goals, Zhongchu Technology actively promotes the integrated development of the industrial chain by focusing on investment, research and development, manufacturing, sales, operation and maintenance services, and recycling related to energy storage battery systems and marine battery systems.

The company's product offerings encompass applications in new energy power generation, smart grids, residential terminals, and green transportation. Zhongchu Technology provides a comprehensive range of energy storage products and integrated solutions that facilitate the convergence of transportation and energy.

4.4.3.8.2 Jiangsu Port Group Co., Ltd.

In 2024, Jiangsu Province proposed to "fully leverage the unique advantages of sea, river, and lake connectivity in Jiangsu, thereby establishing a more distinctive 'Waterway Jiangsu' and achieving a high-quality transition from being a major waterway province to becoming a strong waterway province." In alignment with this objective, the plan aims to significantly advance the deployment of pure electric inland container vessels. By 2025, the province intends to initiate preliminary commercial operations for these vessels, achieve breakthroughs in standards and key technologies related to pure electric transport vessels, and lead the nation in both scale and practical application of inland pure electric transport solutions.

Jiangsu Port Group Co., Ltd., as a large provincial state-owned enterprise backed by state capital, has been entrusted with the responsibility of spearheading national efforts in scaling up and practically applying inland pure electric transport vessels. In collaboration with the Provincial Department of Transport, they have launched China's first 120 TEU pure electric inland container ship featuring a battery swap model—named "Jiangyuan Baishou." This vessel measures 79.92 m in length and 12.66 m in width, capable of reaching speeds up to 17 km/h with an operational range of 220 km. It is powered entirely by batteries and incorporates an innovative plug-and-play battery swap system that allows for battery replacement within just 20 minutes.

4.4.3.8.3 COSCO SHIPPING Group

To implement national major strategies and comprehensively execute the decisions and arrangements of the CPC Central Committee regarding "carbon peaking and carbon neutrality," as well as to adhere to President's directives on "embracing new trends in green, low-carbon, and intelligent shipping development," COSCO SHIPPING Group has developed the world's first 700 TEU pure battery-powered container ship utilizing a battery swap model. This vessel is entirely designed and manufactured in China, featuring 100% domestically sourced equipment that reflects advanced domestic standards.

Focusing on the new energy sector, this project aims to establish a key demonstration vessel for green, low-carbon, and intelligent shipping. It sets several world records concerning length, beam width, container capacity, load tonnage, and battery capacity among electric container ships. The ship measures 119.8 m in length and 23.6 m in width with a maximum speed of 19.4 km/h; it can be equipped with up to 36 mobile containerized batteries as its power source, boasting a total battery capacity exceeding 50,000 kWh.

By employing containerized mobile power sources for battery swaps throughout its journey powered by battery propulsion alone, this innovative design effectively addresses challenges such as prolonged charging times, limited loading capacities, and short cruising ranges associated with traditional electric container ships. Once operational, it is estimated that this single vessel could reduce CO_2 emissions by approximately 2,918 tons annually—equivalent to the yearly emissions produced by about 2,035 passenger cars or the environmental benefit of planting around 160,000 trees each year.

4.4.3.8.4 Hubei Port Group Co., Ltd.

To implement the national "dual carbon goals," promote the green transformation of inland shipping in Hubei Province, and advance the execution of the "Electrification of the Han River" development strategy, Hubei Port Group has undertaken initiatives aligned with provincial government directives. In collaboration with Hubei Donghu Lab, State Grid Hubei Branch, Changjiang Shipping Group Wuhan Ship Design Institute Co., Ltd., and other entities, they have conducted research and executed a pure electric container ship engineering demonstration project—the first of its kind in the Han River Basin.

This initiative represents an effort by Hubei Port Group and its affiliated Central China Port & Navigation Group to capitalize on opportunities within the green and intelligent shipping industry. They have successfully completed construction on "Huaxineng 1," which is recognized as Hubei's inaugural 120 TEU pure electric container demonstration vessel featuring a battery swap model.

The vessel is designed as an open hatch pure battery-driven bulk/container dual-purpose ship, equipped with dual propulsion motors powered entirely by batteries. Measuring 84 m in length, it complies with Class B and C inland waterway standards. The maximum design speed reaches 18 km/h, while offering a single-trip range of up to 175 km—capable of self-sustaining operations for five days.

4.5 Economic Analysis of Battery Swapping Cases

Building upon the sector-specific analyses of battery swap systems for two-wheelers, passenger vehicles, new-energy heavy-duty trucks, vessels, and aircraft presented earlier in this chapter, this section conducts investment feasibility assessments for selected case studies. It should be noted that due to the nascent stage of operational frameworks for maritime and aviation applications, insufficient public data exists to support robust financial modeling. Consequently, our quantitative evaluation focuses exclusively on three commercially mature segments: two-wheelers, passenger cars, and new-energy heavy-duty trucks. Please note that the data sources for our analysis are derived from publicly disclosed reports issued by relevant corporations and market institutions. While certain simplifications have been applied to the revenue and cost calculation methodologies, our assumptions remain prudently conservative. Consequently, the projected outcomes demonstrate greater financial caution relative to certain market benchmarks published by third-party organizations.

4.5.1 Economic Modeling for Two-wheelers

Based on the descriptions above, taking Yugu Technology as a case study, an initial return calculation for its battery swap project can be conducted.

In terms of capital investment, we base our calculations on yuan 110,000 allocated per individual battery swap cabinet. According to official disclosures, as of June 2023, Yugu Technology had deployed 23,700 battery swap cabinets across 66 Chinese cities, supported by 517,500 batteries serving 391,100 registered users. This translates to approximately 16 users per cabinet on average.

Market analysis (Frost & Sullivan) identifies on-demand delivery personnel as the primary battery swap service adopters (2023 baseline). Concurrent with domestic instant delivery market expansion and supporting infrastructure enhancement, two-wheeler battery swap service penetration exhibits significant growth potential. Quantitative modeling forecasts sector expansion from yuan 7.08 billion (2023) to yuan 20.11 billion (2026) at CAGR of 41.62% (Figure 4.35).

Consequently, user base expansion remains achievable, yet China's delivery rider population has passed its hyper-growth phase in recent years. Projecting a 40% CAGR is therefore not deemed reasonable. We consequently model a 20% CAGR for the user base over the next decade. Assuming the average user count per battery swap cabinet also grows at an annual rate of 20% over the next decade, with calculations based on: two times daily swaps per vehicle owner; each swapped battery at fully charged state; single battery capacity is 1.2 kWh. Based on market research

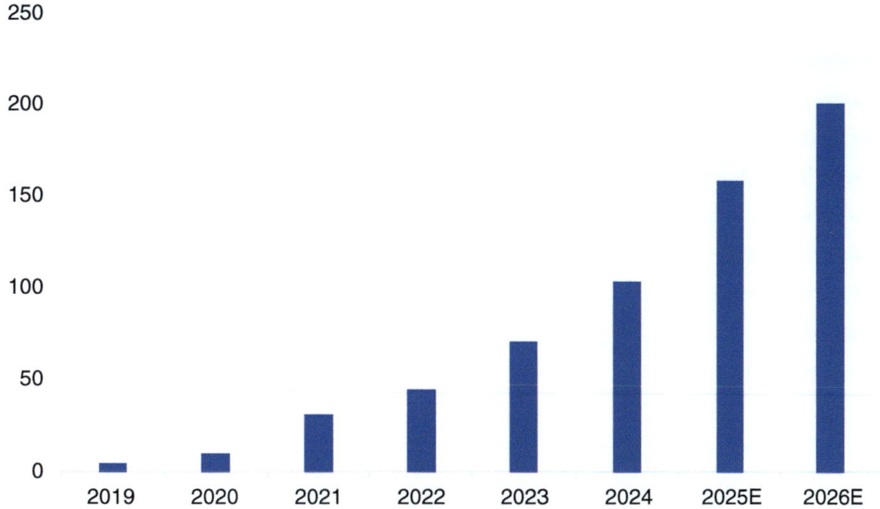

Figure 4.35 Market size of two-wheel battery swap. *Source:* Frost & Sullivan.

findings that the composite electricity price (including service fee) is yuan 1.6/kWh for peak-hour charging of two-wheeled EVs, we calculate the revenue projections per battery swap cabinet as follows[3] (Table 4.7):

Cost components include electricity procurement, equipment maintenance and labor cost. EV battery swap stations concentrate electricity procurement during off-peak periods. In recent years, China's off-peak electricity rates have remained stable at approximately yuan 0.35/kWh.[4] Based on previously estimated daily energy transfer volumes, procurement costs can be calculated. Labor costs represent wages for operational staff, allocated per swap cabinet at 4% of revenue.[5] Maintenance fees align with consensus forecasts across market institutions, consistently maintained at 2% of total capital investment. The detailed cost projections are structured as follows (Table 4.8).

Based on an investment cost of ¥110,000 per swap cabinet, we calculate a payback period of 4.82 years and an internal rate of return (IRR) of 6% using projected annual net cash inflows.

4.5.2 Economic Modeling for Vehicles: A NIO Case Study

As established above, NIO pioneered the battery swap model and has progressively achieved operational maturity, with economies of scale now materializing. As the representative enterprise for passenger vehicle battery swapping, this section conducts revenue modeling using NIO as the benchmark. Unlike two-wheeled battery swap cabinets, passenger vehicle swapping exhibits lower per-vehicle swap frequency; therefore, economic assessments must be comprehensively modeled at the integrated charging facility level.

Table 4.7 Annual analysis of battery swap cabinet operation data.

Year	1	3	5	7	9
Persons/cabinet	16	23	33	48	69
Swap frequency/day	32	46	66	96	138
Electricity/charge (kWh)	1.2	1.2	1.2	1.2	1.2
Electricity/day (kWh)	38.40	55.30	79.63	114.66	165.11
Electricity price	1.6	1.6	1.6	1.6	1.6
Electricity revenue/day	61.44	88.47	127.40	183.46	264.18
Days/year	365	365	365	365	365
Total revenue	22,426	32,293	46,502	66,962	96,426

3 Based on the industrial and commercial electricity prices in Shanghai.
4 Based on the industrial and commercial electricity prices in Shanghai.
5 Based on broker research reports, including those from Cinda Securities and Tianfeng Securities.

Table 4.8 Annual cost and net cash flow analysis.

Year	1	3	5	7	9
Power procurement price (yuan/kWh)	0.35	0.35	0.35	0.35	0.35
Annual procurement cost	4,906	7,064	10,172	14,648	21,093
Annual maintenance fees	2,200	2,200	2,200	2,200	2,200
Annual labor cost	449	646	930	1,339	1,929
Annual total cost	7,554	9,910	13,302	18,187	25,222
Net cash inflow	14,871	22,383	33,199	48,775	71,204

According to NIO's official operational disclosures, the company recorded a daily average of 78,000 battery swap sessions throughout 2024. With an operational portfolio of 2,800 swap stations, this yields a derived daily average of 28 swap sessions per station.

The deployment cost per battery swap station is yuan 3.5 million, primarily attributed to battery inventory and swapping equipment. According to NIO 's strategy, the company plans to deploy 3,600 operational stations by end-2025, expanding to over 4,000 stations in 2026, marking the completion of its recent expansion cycle. Consequently, the forecast model assumes 3,600 stations in Year 1 (2025) and 4,000 stations in Year 2 (2026), maintaining this stabilized capacity throughout the subsequent projection horizon.[6]

Projections for the daily swaps per station are based on the growth rate of swap stations. Using the baseline daily swap volume of 28 swaps/station/day, and accounting for station expansion to 3,600 stations in Year 1 and 4,000 stations in Year 2, the model calculates a daily swap volume per station of 36 in Year 1 and 40 in Year 2. Subsequent periods stabilize at equilibrium.

The electricity pricing assumptions align with the methodology established in the two-wheeler battery swap section: a swap service fee of yuan 1.6/kWh and a procurement cost of yuan 0.35/kWh. Given an EV battery capacity of 75 kWh per pack, we project station-level revenue as follows (Table 4.9).

Regarding operational costs beyond electricity procurement, battery swap stations incur maintenance expenses, labor costs, and facility rentals. A standard station occupies four parking spaces, with an annual rental cost of 50,000 yuan per space, resulting in a total rental expenditure of 200,000 yuan. Maintenance fees are calculated at 5% of the capital investment, amounting to 150,000 yuan annually. Labor costs are projected at 100,000 yuan per year. Consequently, the total annual operating costs sum to 450,000 yuan. Total cost calculation is shown in Table 4.10.

[6] Source: NIO's official website and public interviews.

Table 4.9 Annual operation and revenue analysis of battery swap stations.

Year	1	2	3	4	5
Daily swap sessions per station	36	40	40	40	40
Rated energy per battery pack (kWh)	75	75	75	75	75
Daily energy dispensed per station (kWh)	2,200	2,200	2,200	2,200	2,200
Electricity retail price (yuan/kWh)	1.6	1.6	1.6	1.6	1.6
Days per year	365	365	365	365	365
Annual revenue (yuan)	1,576,800	1,752,000	1,752,000	1,752,000	1,752,000

Table 4.10 Annual cost and net cash flow analysis of battery swap stations.

Year	1	2	3	4	5
Annual electricity procurement cost	551,880	613,200	613,200	613,200	613,200
Annual rental expense	200,000	200,000	200,000	200,000	200,000
Annual maintenance fees	150,000	150,000	150,000	150,000	150,000
Annual labor cost	100,000	100,000	100,000	100,000	100,000
Annual total cost	1,001,880	1,063,200	1,063,200	1,063,200	1,063,200
Net cash flow	574,920	688,800	688,800	688,800	688,800

Based on the foregoing analysis, the payback period for a single automotive battery swap station is calculated at 5.25 years. It should be noted that this case study utilizes NIO's official 2024 operational data—a period when its battery swap model had been operational for a significant duration. Consequently, the projected returns appear robust. However, multiple market institutions concur that NIO is expected to achieve full payback on its initial swap infrastructure investments within the next three years.

4.5.3 Economic Modeling for Heavy Trucks

Heavy-duty truck battery swap stations require significantly higher swap frequency and capital investment compared to passenger vehicle stations, though both follow fundamentally similar financial evaluation frameworks. Investment analysis reveals cost concentration in swap equipment and battery assets. According to disclosures by GCL Energy Technology and HNC Technology, per-station equipment costs range from 2 to 4 million yuan, yielding a median equipment investment of 3 million yuan. Operationally, a standard station necessitates 7 backup battery packs, while heavy-duty stations demand an additional 40 in-service packs. Battery investments thus total 15 million yuan per corporate filing. Industry benchmarks indicate 2.3 million yuan for grid connection and auxiliary investments. Consequently, the total capital expenditure for a single heavy-duty swap station reaches 20.3 million yuan.

Regarding the average daily swap frequency for heavy-duty truck battery swap stations, industry corporate disclosures and market data indicate a daily average of 50 swap cycles per station. According to announcements by China's Ministry of Industry and Information Technology (MIIT), the typical battery capacity for new-energy heavy-duty trucks in the market as of early 2025 is approximately 400 kWh. Therefore, assuming each full-capacity swap delivers 400 kWh, the annual energy throughput reaches 7,000 MWh.[7]

Regarding electricity pricing, we retain the previously established sales rate of 1.6 per yuan kWh. Based on this, the station's revenue can be calculated as follows (Table 4.11).

The cost structure remains consistent with prior analysis. Each heavy-duty swap station requires a crew of four operators. Given the higher operational intensity

Table 4.11 Annual operation metrics and revenue analysis of battery swap stations.

Year	1	2	3	4	5
Daily swap sessions per station	50	50	50	50	50
Rated energy per battery pack (kWh)	400	400	400	400	400
Daily energy dispensed per station (kWh)	20,000	20,000	20,000	20,000	20,000
Electricity retail price (Yuan/kWh)	1.6	1.6	1.6	1.6	1.6
Days per year	350	350	350	350	350
Annual revenue (thousand Yuan)	11,200	11,200	11,200	11,200	11,200

[7] Source: SWS Research.

Table 4.12 Annual cost and net cash flow forecast for battery swap stations.

Year	1	2	3	4	5
Annual electricity procurement cost	2,450,000	2,450,000	2,450,000	2,450,000	2,450,000
Annual rental expense	200,000	200,000	200,000	200,000	200,000
Annual maintenance fees	1,015,000	1,015,000	1,015,000	1,015,000	1,015,000
Annual labor cost	400,000	400,000	400,000	400,000	400,000
Annual total cost	4,065,000	4,065,000	4,065,000	4,065,000	4,065,000
Net cash flow	7,135,000	7,135,000	7,135,000	7,135,000	7,135,000

driven by elevated swap frequency, the annual salary per operator is calculated at 100,000 yuan. Maintenance costs, similarly, influenced by utilization rates, are projected at 5% of the total capital investment. Electricity procurement expenses maintain the previously established rate of 0.35 yuan/kWh. The resultant cost breakdown is detailed in Table 4.12.

Based on the aforementioned analysis, the investment payback period for heavy-duty truck battery swap stations is calculated at 2.85 years. This metric demonstrates a shorter capital recovery cycle compared to two-wheeler swap cabinets and passenger vehicle stations, indicating superior profitability in the heavy-duty segment's business model.

5

Overview of the Battery Bank Industry

5.1 Introduction to Battery Bank Industry

In the construction costs of battery swapping stations, apart from the capital expenditure on the stations themselves, the cost of batteries also constitutes a significant portion. Currently in China, early-generation new energy vehicles and their batteries are experiencing a wave of retirement. Simultaneously, with the continuous development of the battery swapping model that separates vehicles from batteries, a new concept—the "battery bank"—has arisen in response. This chapter will introduce this industry, which is a new concept emerging in the new energy industry chain and an important part of it.

5.1.1 Political Background and Environment

As a pioneer in the new energy vehicle market with early adoption and rapid growth, China is now witnessing the first wave of retirement for its earliest generation of electric vehicles (EVs). With this retirement wave, power batteries—the core assets of these vehicles—are also entering a phase of mass decommissioning. The retirement of power batteries carries significant environmental implications. Containing heavy metals like nickel, cobalt, manganese, and lithium, along with organic substances such as electrolytes, these batteries pose severe threats to soil and water sources when leaked into the environment, disrupting ecological balance and endangering human health. Thus, the Chinese government places strong emphasis on promoting the recycling and reuse of battery assets. The Chinese government has begun rolling out policies to encourage the battery swapping model (vehicle-battery separation) and the recycling of battery assets early.

A compilation of related Chinese government policies has been summarized in Table 5.1.

Besides China, both Europe and the United States have enacted legislation in the new energy battery sector. In Europe, the landmark regulation is the new battery

Table 5.1 Key policies and initiatives for battery swapping and recycling in China (2019–2025).

Date	File	Core content
June 2019	Implementation Plan for Promoting the Upgrading of Key Consumer Products and Facilitating Resource Recycling	Promote battery-leasing services and other battery-swapping models for new energy vehicles; encourage the development of NEVs compatible with both charging and battery-swapping technologies.
April 2021	Notice on Launching the Application Pilot Program for New Energy Vehicle Battery-Swapping Models	Encourage private capital participation in establishing battery asset management companies ("battery banks")
March 2023	Official Announcement Regarding Product Certification Implementation for Cascaded Use of Retired NEV Batteries	Improve the market ecosystem for cascaded utilization of power batteries and foster the healthy and well-regulated development of the battery echelon-reuse industry
May 2023	National Catalogue of Advanced and Applicable Process Technologies and Equipment for Comprehensive Industrial Resource Utilization (2023 Edition) (Draft for Comment)	Include the following in the catalog: 1. Automated disassembly of NEV power battery cells with cathode/anode material reclamation technology 2. Full-process high-value utilization technology and equipment for end-of-life power batteries
December 2024	Industry Standard Specifications for Comprehensive Utilization of End-of-Life Power Batteries from New Energy Vehicles (2024 Edition)	Establish annual thresholds for cascaded utilization of end-of-life power batteries, with enhanced requirements for advancing R&D and deployment of recycling technologies, equipment, and processes targeting battery components including cathode/anode materials, separators, and electrolytes.
February 2025	Action Plan to Strengthen the Ecosystem for Recycling and Reuse of New Energy Vehicle Power Batteries	Emphasize full lifecycle management, digital traceability, and standardization framework development for power batteries to promote industry-wide compliance

Source: Ministry of Industry and Information Technology of the People's Republic of China.

and waste battery regulation (commonly known as the new battery regulation), whose core requirements include:

1. **Carbon footprint declaration system:** Requiring electric vehicle batteries, light transport batteries, and industrial batteries above 2 kWh to disclose full lifecycle carbon data.

2. **Battery passport system:** Mandating that such batteries feature a "Digital Battery Passport" from January 2026, recording carbon footprint, material composition, and supply chain information, with traceability enabled via QR codes throughout their lifecycle.
3. **Recovery rate targets:** Gradually increasing minimum recycling rates of critical metals, cobalt, lead, lithium, and nickel, in batteries, targeting >95% for cobalt, lead, nickel, and copper and >80% for lithium by 2031.

The United States has also enacted new energy legislation, though this act prioritizes promoting domestic industries. Specifically, the Inflation Reduction Act (IRA) launched in August 2022 provides up to USD 7,500 in tax credits for purchasing qualifying new energy vehicles that meet critical mineral and battery component requirements. The critical mineral mandate requires that lithium, cobalt, nickel, and other key minerals in batteries be extracted, processed, or recycled in North America or free-trade-agreement partner countries, with the proportion increasing progressively from 40% in 2023 to 80% by 2027. The battery component rule stipulates that the percentage of battery components manufactured or assembled in North America must rise from 50% in 2023 to 100% by 2029.

In summary, as the most mature market for new energy vehicles, China has established extensive supportive legislation that creates a robust legal framework for battery bank business models. Europe's New Battery Regulation similarly bolsters industries involved in power battery circularity. While US policies prioritize domestic industry protection, they still offer subsidy-driven growth prospects for relevant sectors.

5.1.2 Definition and Business Model

The battery bank is a new type of battery asset management business model developed based on the "vehicle-battery separation" model. Its core lies in separating the power battery from the fixed assets of the vehicle and transforming it into a tradable and manageable "digital asset," and achieving the maximization of full life cycle value through professional operation. In the business model of the battery bank, the participants include battery asset management and utilization companies (battery banks), the government, battery manufacturers, vehicle manufacturers, energy storage enterprises, battery recycling plants, charging and swapping operators, and insurance companies, as illustrated in Figure 5.1. The government, battery suppliers, vehicle manufacturers, swapping operators, and financial institutions jointly invest to establish a special fund for battery power, and the special fund establishes a battery management company to provide battery leasing services to users. It also monitors the battery usage status of users in real time and promptly makes operational scheduling and necessary health management measures. Each participating entity has its own business model, and they all seek what they need to jointly achieve a win–win outcome.

For the government, a swappable battery is a kind of public asset for society, with infrastructure characteristics, guiding the orderly development of the electric battery swapping industry can help create more social benefits. For automobile

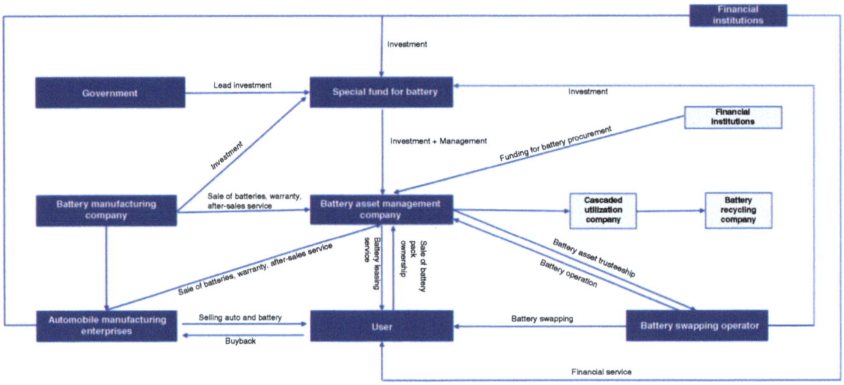

Figure 5.1 Business model of battery bank industry.

manufacturing companies, the business model of the battery bank can help reduce vehicle prices, diversify sales methods, facilitate promotions, facilitate battery monitoring, and prevent battery failures from causing recall issues. For a battery manufacturing company, it can increase battery sales, strengthen monitoring of the battery's entire life cycle, and facilitate battery recycling and secondary utilization. For battery swapping operators, they can earn service fees for battery swapping and the difference between electricity charges, and provide additional services through the battery bank. For financial institutions, by providing loans and launching relevant financial derivatives, they can obtain corresponding benefits. Currently, the types of financial derivatives in the battery bank industry are relatively few, and there is considerable room for development.

Currently, the battery bank's revenue model comprises three primary streams: power battery management operations and leasing, cascaded utilization and recycling, and financial derivatives provision (e.g., insurance). Battery leasing constitutes the core revenue source. Should battery-leasing models achieve widespread adoption, consumers purchasing EVs would automatically enter into bundled battery lease agreements. In this scenario, battery banks could determine rental fees based on battery capacity, performance metrics, and lease duration, with the EV itself serving as collateral for rental payments if necessary. Through large-scale leasing operations, battery banks gain stable cash flows while strategically aligning lease terms with projected battery lifespans—enabling circular asset management. Subsequently, when retired EV batteries retain sufficient capacity for less demanding applications, battery banks may sell them to energy storage systems, low-speed EVs, or power tool manufacturers for asset disposal revenue and provide value-added services for cascaded-use products, generating service fees.

Insurance also plays an indispensable role in the battery bank model by providing both risk protection and value creation. On one hand, while the battery bank model maintains the fundamental purpose of property insurance—protecting relevant parties from losses—it shifts policyholders from individual consumers to battery banks themselves. On the other hand, the "battery bank + insurance" business model is poised to facilitate a transformation in how the insurance industry creates value within the new energy sector. In conclusion, the high compatibility between insurance institutions and the battery bank model stems from insurance's unique attributes precisely addressing the gaps in the battery bank's asset risk control. Such cross-industry convergence generates synergistic effects that achieve exceptional outcomes unattainable by either battery banks or insurers operating independently. This injects robust and sustainable momentum into the new energy vehicle industry ecosystem.

Beyond the three traditional profit models mentioned above, data monetization emerges as a future growth driver for battery banks, given their extensive battery resources and associated data. By establishing operational, algorithmic, and data infrastructure, these institutions can implement real-time monitoring and dispatch throughout batteries' entire life cycle. This transforms data itself into monetizable assets, delivering actionable insights to battery suppliers, automakers, and battery swap operators.

5.2 Industry Analysis

5.2.1 Forecast of the Battery Bank Industry Market Size

The market size analysis and projection for the battery bank sector can be segmented into three core revenue streams, as outlined in Section 3.1.2: battery leasing, battery recycling and grading/reuse, and battery insurance. Accordingly, this chapter conducts quantitative projections for each segment across both Chinese and global markets.

5.2.1.1 China

From a demand perspective, the battery leasing segment demonstrates substantial revenue scale and rapid growth, primarily driven by the accelerating penetration of battery swappable vehicles over the next 5–7 years. Recent policy tailwinds for battery swapping solutions and aggressive automaker engagement are catalyzing this expansion. It is projected that Annual sales of battery swappable passenger and light-duty logistics vehicles will reach about 2.5 million units in 2025, and sales will exceed 9 million units in 2030, as can be read in Figure 5.2. The penetration rate would exceed 50% of total NEV sales (passenger + light logistics vehicles). As swappable vehicle sales surge and end-user acceptance of battery leasing models intensifies, this segment is poised for massive expansion. The battery leasing market is projected to reach 67.8 billion yuan ($\approx$9.5 billion USD) by 2030, which is illustrated in Figure 5.3.

On the other hand, as the battery insurance market matures, its scale is expected to grow rapidly over the next five to seven years, with premiums projected to exceed 20 billion yuan by 2030, as shown in Figure 5.4. The growth of the battery insurance market will be primarily driven by the expansion of battery swap-enabled vehicles under the battery bank model and increased demand for battery asset risk management. Currently, after securing centralized insurance for batteries, battery

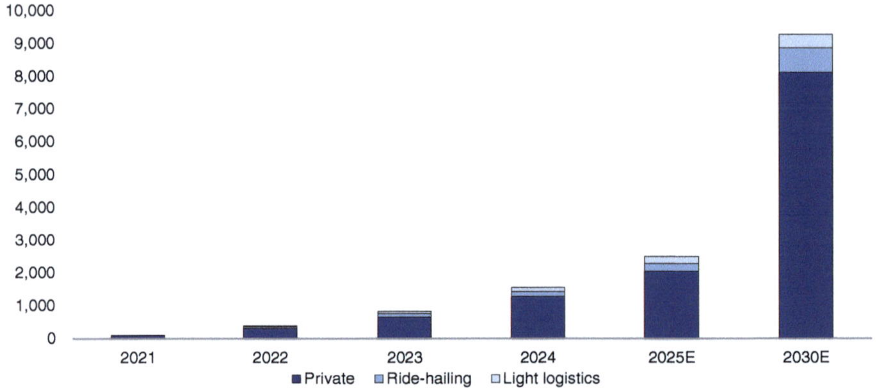

Figure 5.2 China battery swappable passenger and light duty logistics vehicles projection (in thousands units). *Source:* From the research of Strategy&, PwC's strategy consulting business.

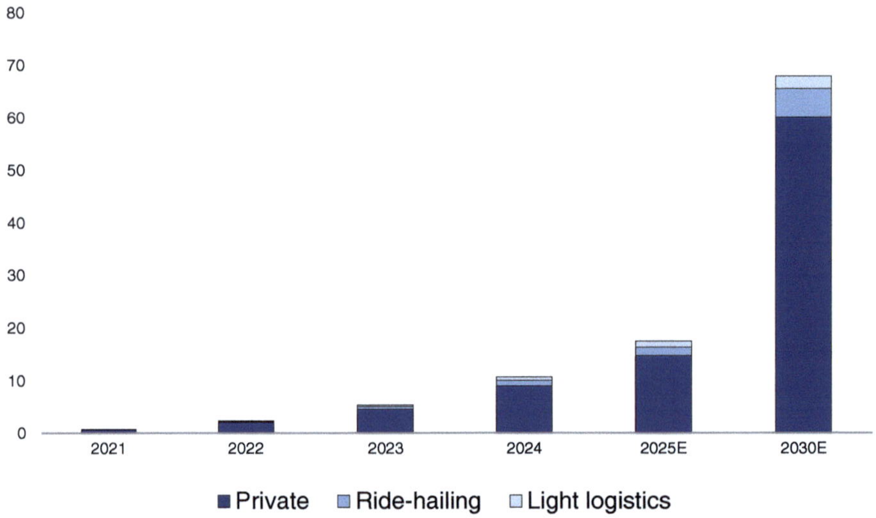

Figure 5.3 Market size forecast for battery leasing of battery swappable passenger vehicles and light logistics vehicles in China (in yuan billions). *Source:* From the research of Strategy&, PwC's strategy consulting business.

banks will bundle insurance fees into battery lease subscriptions to generate premium income. In the future, coverage may extend to battery swap users who opt out of subscription plans. Furthermore, deeper collaboration with insurers and regulatory developments could create additional revenue streams for battery banks through second-life applications (e.g., energy storage) and recycled materials.

Thirdly, the market potential for battery second-life applications and recycling is poised to unlock, growing into a hundred-billion-yuan industry, as presented in Figure 5.5. Battery banks, leveraging their insights into battery data and scale advantages in asset ownership, are positioned to gain significant influence while

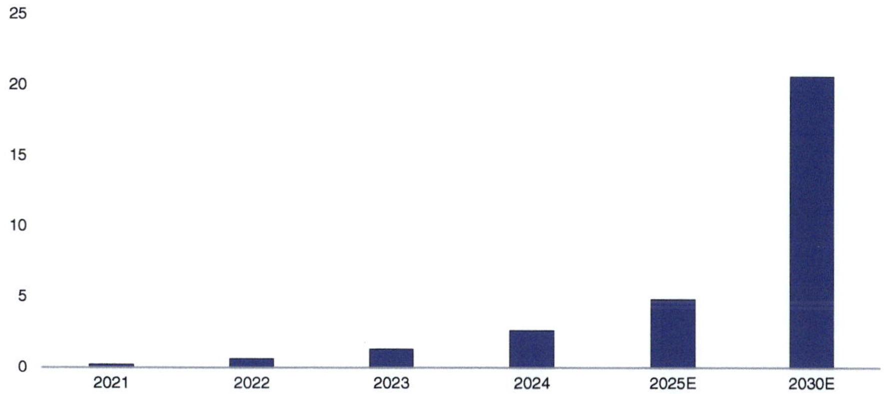

Figure 5.4 Market size forecast for insurance premium of battery swappable passenger vehicles and light logistics vehicles in China (in billions of yuan). *Source:* From the research of Strategy&, PwC's strategy consulting business.

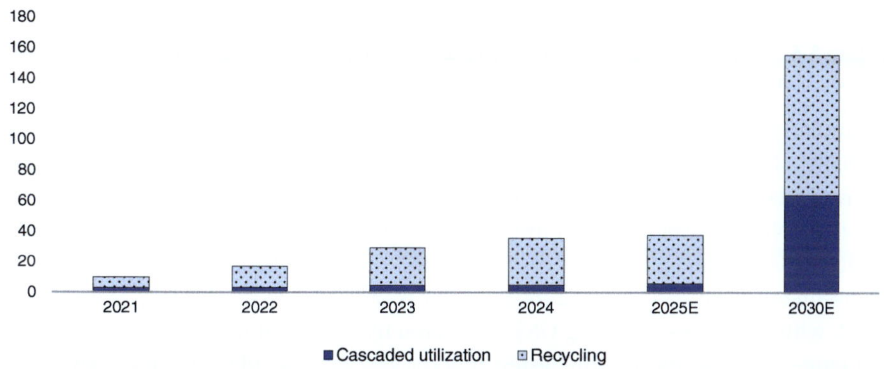

Figure 5.5 Market size forecast for battery cascaded utilization and recycling market scale in China (in billions of yuan). *Source:* From the research of Strategy&, PwC's strategy consulting business.

capturing substantial value in this market. Should prices of battery raw materials stabilize after historical volatility, the second-life and recycling sector will accelerate rapidly. China's market is projected to reach 155.3 billion yuan by 2030, with second-life applications accounting for 63.7 billion yuan and battery recycling 91.6 billion yuan.

5.2.1.2 Global

Globally, as the battery bank business model remains nascent and has yet to gain traction outside China, the initial market scale will be dominated by the Chinese market. With the increasing penetration of new energy and battery swap-enabled vehicles worldwide, the market is poised for gradual expansion. As can be seen from Figure 5.6, the global battery leasing market is projected to reach US$176.1 million in 2024, growing at a 22.4% CAGR from 2025 to 2034, and is expected to hit US$1.33 billion by 2034.

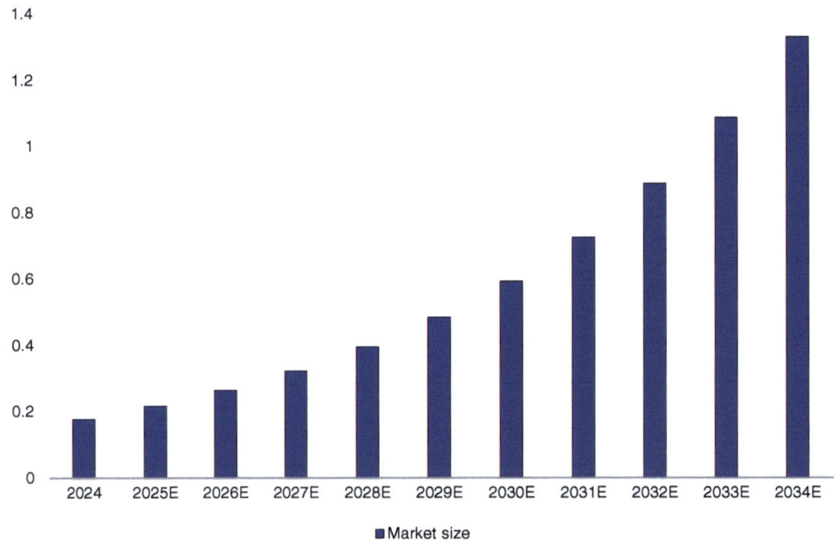

Figure 5.6 Market size forecast for battery leasing in global (in USD billions). *Source:* Global Market Insights.

Similarly, due to varying adoption rates of new energy and battery bank models globally, standalone market data for the global battery insurance sector remains unavailable. QYResearch's study extrapolates the global battery insurance scale based on total NEV insurance premiums. Empirical case studies indicate that battery-related coverage constitutes approximately 40% of comprehensive insurance policies. Consequently, the battery insurance market is projected to reach US$770 million in 2024, with forecasts surpassing US$1.53 billion by 2029, as illustrated in Figure 5.7.

Figure 5.8 presents the forecasted market size for cascade utilization and recycling. Globally, the battery recycling and cascaded utilization market is projected to reach US$6.88 billion in 2024. Driven by increasing adoption rates of electric vehicles worldwide, the volume of decommissioned batteries requiring recycling will surge, pushing the market size beyond US$70 billion by 2030—representing a staggering 50% CAGR.

In summary, China has now entered a surge of decommissioned batteries, driving exponential growth in its battery bank market. While China currently possesses the world's most mature new energy ecosystem, this demonstrates compelling growth potential for other regions—particularly Europe and the United States. As EV adoption accelerates globally, substantial expansion is anticipated across three key sectors: battery leasing, battery insurance, and battery recycling/cascaded utilization markets.

5.2.2 Competitive Landscape in the Battery Bank Sector

As established, the battery bank business model is currently implemented primarily in China. This section will therefore analyze the competitive landscape of China's battery bank market. Presently, diverse enterprises and capital entities are scrambling to position themselves in this sector, which can be categorized into three

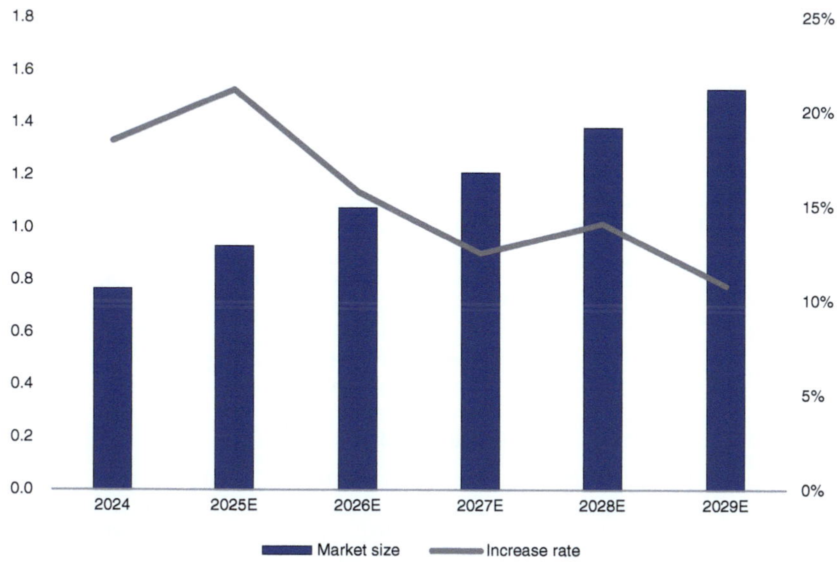

Figure 5.7 Market size forecast for insurance premium in global (in USD billions). *Source:* QYResearch.

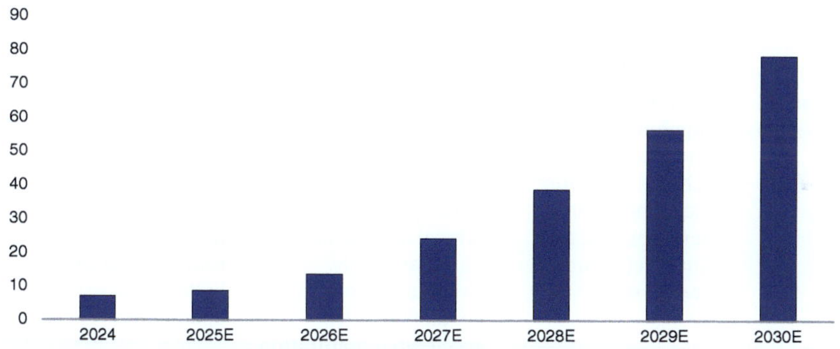

Figure 5.8 Market size forecast for battery cascaded utilization and recycling market scale in global (in USD billions). *Source:* Market Monitor.

dominant models: OEM-led, battery manufacturer-led, and battery swap service provider-led. Given that battery asset scale and lifecycle value optimization constitute critical competitive edges, battery cell manufacturer-led models may secure competitive advantages through superior battery compatibility and integrated industrial chain presence. Table 5.2 presents a comprehensive comparative analysis of the three competitive models.

5.2.3 Future Development Trend in the Battery Bank Sector

As noted earlier, the battery banking model remains in a nascent phase with evolving business architectures, where numerous operational frameworks require further maturation. Given its attractive prospects, this section will forecast industry

Table 5.2 Strategic comparison of battery swap models: battery makers vs. OEMs.

	Battery manufacturer	OEM-led	Battery swap service
Representative players	EVOGO	ENERGIEX	GCL Energy Technology Co., Ltd.
Strategic positioning imperatives	Boost Battery Product Volume and Diversify Revenue Streams.	Capture emerging business opportunities from the battery swapping model, where controlling battery assets and related data assets delivers both financial appeal and strategic leverage for auto OEMs.	Establish a closed-loop battery swap ecosystem to enhance synergistic value and amplify capital markets appeal.
Strengths	Highly standardized battery packs unlock greater potential for multi-OEM collaboration, facilitating battery asset scale expansion and customer portfolio optimization.	Streamlined battery asset procurement with secured supply continuity.	Improve the market ecosystem for cascaded utilization of power batteries and foster the healthy and well-regulated development of the battery echelon-reuse industry.
Weakness	Current capabilities remain underdeveloped in echelon utilization, recycling chain development, and financial-insurance partnerships.	Limited interoperability due to competitive barriers restricts multi-brand participation, hindering critical asset mass. Simultaneously, circular economy enablers require accelerated maturation.	Battery asset portfolios may struggle to achieve economies of scale, potentially driving more automakers toward proprietary battery-as-a-service operations.

evolution trajectories and emergent opportunities, along with conducting cross-impact analysis relative to the charging/swapping ecosystem.

5.2.3.1 Standardization Gaps

The foremost challenge facing battery banking is the lack of standardization, where incompatible battery specifications across vehicle models cause swapping failures. Competing IP protection agendas among battery and vehicle manufacturers further impede unification. According to China's MIIT Announcement on Road Motor Vehicle Manufacturers and Products, the 2019 edition documented 98 enterprises with 307 new energy vehicle model entries, which include 279 battery electric vehicle (BEV) models from 92 manufacturers, 11 plug-in hybrid electric vehicle (PHEV) models from 9 manufacturers, and 17 fuel cell electric vehicle (FCEV) models from 8 manufacturers.

Regarding standardization, continuous advancement of battery banking will drive industry-wide regulatory frameworks for unified protocols. Market consolidation through competitive selection is poised to establish dominant OEM alliances, accelerating battery and platform standardization. Battery banking demonstrates stronger value propositions for ride-hailing fleets through vehicle homogenization, whereas private owners with lower annual mileage (typically <20,000 km) face limited economic incentives. A novel solution involves splitting the battery into two modular units—replacing a single large battery with two smaller ones. One serves as a fixed standard pack, while the other functions as a swappable optional module. Vehicles can be factory-equipped with a base battery (0–200 km range) for daily commutes. For occasional long trips or urgent needs, users can temporarily lease an additional range-extending battery. This hybrid energy replenishment model—partial charging combined with partial swapping—maintains total driving range while enabling flexible battery capacity and mileage customization (Jingbo Li, 2020).

5.2.3.2 Data Monetization

As discussed in Section 3.1.2 (Chapter 3), constrained by current market penetration, the monetization of battery data assets remains premature. However, scaling battery banking operations will exponentially expand datasets, leveraging AI algorithms to forecast battery health status, while implementing blockchain for immutable data verification will transform battery datasets into financially valuable assets, thereby unlocking innovative business paradigms. The concept is illustrated in Figure 5.9.

Tesla's battery management system (BMS) utilizes smart algorithms and AI for real-time battery data monitoring, while Huawei's AI-BMS has been deployed in select EV models. Continuous refinement of AI-powered battery systems will unlock exponential growth in data asset valuation.

5.2.3.3 Virtual Power Plant

A virtual power plant (VPP) is an energy management system that integrates distributed energy resources (DER)—such as photovoltaic, energy storage, Ev, and demand response loads—via communication and control technologies. It enables unified operational coordination and market participation, with core value propositions in flexible regulation, aggregated response, and intelligent trading. As battery banking scales, demand for battery swap stations will rise significantly. Integrating these stations with advanced metering infrastructure enables electricity load shifting, unlocking economic benefits through peak shaving, valley filling, and participation in wholesale energy markets.

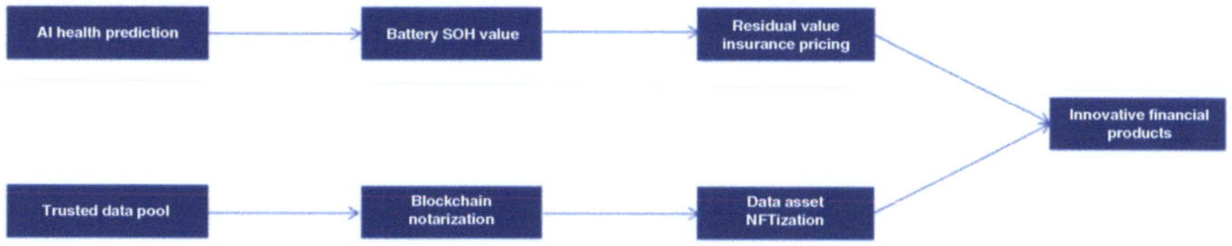

Figure 5.9 Battery bank data monetization process.

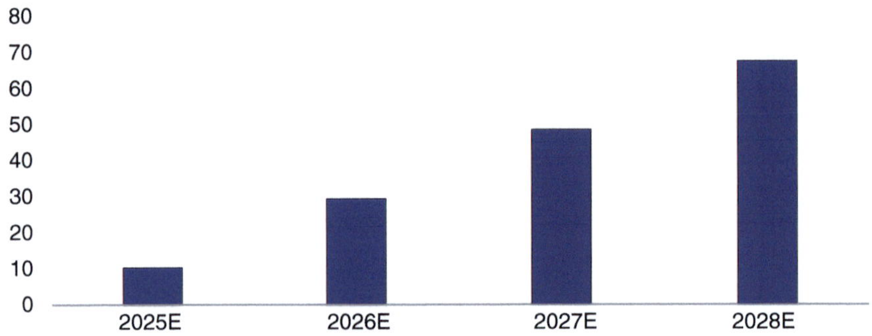

Figure 5.10 China VPP market size (in billions of yuan). *Source:* ASKCI Consulting Co., Ltd.

The VPP concept has reached maturity in Europe and the United States, with future growth potential concentrated in China and emerging markets. In the United States, Tesla has established robust VPP capabilities centered on its Auto-bidder real-time trading and control platform. Meanwhile in Europe, Germany's Next Kraftwerke operates as the continent's largest VPP provider, serving dual roles as an energy trader and grid-balancing aggregator.

China's VPP sector is transitioning from policy-driven pilots to scaled commercialization, with substantial growth potential projected. The market is forecasted to reach 10.2 billion yuan by 2025, and is further expected to surpass the 100-billion-yuan threshold by 2030, as illustrated in Figure 5.10.

5.2.3.4 Carbon Footprint

As we discussed in Section 3.1.1 (Chapter 3), the EU *Batteries and Waste Batteries Regulation* mandates carbon footprint declaration, labeling, and disclosure, while China's carbon accounting framework is rapidly evolving. Battery banking platforms, as large-scale battery asset holders, stand to integrate carbon footprint metrics into their business models, unlocking synergistic value through carbon labeling mechanisms.

Carbon footprint management necessitates the quantified calculation of greenhouse gas emissions. By leveraging digitized carbon metrics, enterprises enhance brand sustainability profiles while aligning with global decarbonization imperatives. Future development of dedicated carbon management platforms for battery banking assets will streamline emission tracking, transforming compliance into economic value. As carbon regulations mature worldwide, these footprints evolve into monetizable assets within expanding carbon markets—generating environmental and financial returns in concert.

6

Key Players and Strategic Positions in the Battery Swapping Ecosystem

The battery swapping industry chain encompasses numerous components and can be categorized into three primary segments based on the "manufacturing-operation-user" framework: upstream, midstream, and downstream.

Upstream: This segment includes battery suppliers, basic component providers for battery swapping stations, and suppliers of supporting charging systems. These entities are responsible for delivering appropriate batteries and equipment for battery swapping stations, as well as software systems tailored to specific applications. Following several rounds of technological advancements, most manufacturers are now equipped to supply comprehensive sets of equipment for battery swapping stations.

Midstream: This segment is primarily composed of manufacturers of complete vehicles (including ships and aircraft), operators, and centralized battery banks. Vehicle manufacturers and operators play a crucial role in establishing and managing battery swapping stations while providing related services to the market. Battery banks oversee the management and leasing of batteries centrally, thereby creating an ecosystem that fosters mutual benefits among consumers, new energy vehicle enterprises, and the banks themselves.

Downstream: This segment consists of users who utilize battery swapping services along with entities involved in battery recycling. Presently, there is a heightened demand for battery swapping models from commercial vehicles, taxis, and other B-end users.

An illustrative battery swap process diagram for different applications is presented in Figure 6.1.

At this stage in its development, the mature battery swapping industry predominantly serves passenger cars as well as commercial vehicles. Notable Chinese operators within this sector include:

New energy vehicle manufacturers: These companies establish their own network of battery swapping stations designed to complement their new energy vehicle offerings; prominent examples include NIO, Geely, and First Technology.

Third-party operators include GCL Energy, Aulton New Energy, and Qiyuan Green Power. Major battery suppliers consist of CATL, BYD, and EVE Energy. Providers of charging systems encompass XJ Electric, Tonhe Electronic Technology, JONHON, and Yonggui Electric Equipment. Notable suppliers of battery swapping systems are Weida Group, CSG Smart Science & Technology, Harmontronics Automation Technology, and BOZHON Precision Industry Technology.

Furthermore, the battery swapping industry for electric ships and the low-altitude economy is in its nascent stages; thus, there remains significant potential for growth within this sector.

An overview of battery swap industry chain and its key stakeholders is illustrated in Figure 6.2.

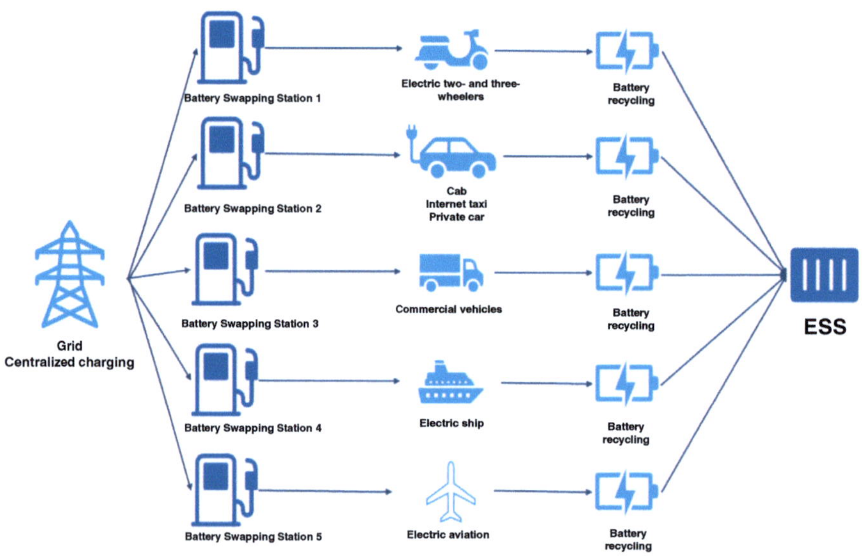

Figure 6.1 Battery swap process diagram for different applications.

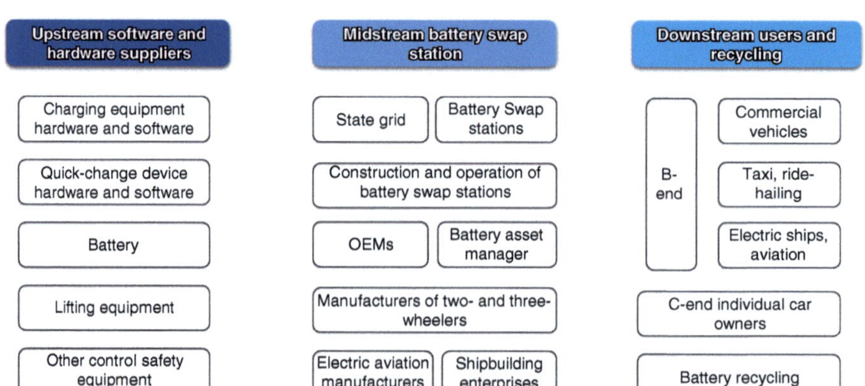

Figure 6.2 Battery swap industry chain structure diagram.

6.1 Upstream Hardware and Software Suppliers

6.1.1 Battery Cell Manufacturers

As a leading enterprise in the battery industry, CATL plays a pivotal role in the research, development, and application of battery swapping technologies. Below is a comprehensive introduction to CATL's battery swapping products tailored for various usage scenarios, including passenger vehicles, light trucks, pickups, and micro-trucks within the logistics sector:

6.1.1.1 Planning of Battery Swapping Products for Different Scenarios

1. **Battery swapping products for passenger cars**
   In response to the growing demands of passenger car users for enhanced convenience and efficiency, CATL has developed specialized battery swapping solutions tailored specifically for passenger vehicles. These products emphasize a compact design to accommodate the limited spatial configurations typical of passenger cars. Simultaneously, through the optimization of battery architecture and management systems, rapid battery swapping is facilitated, significantly reducing user waiting times.

   For instance, advanced battery technologies are employed to enhance energy density, thereby ensuring an extended driving range without increasing the overall volume or weight of the batteries. Furthermore, CATL collaborates closely with automobile manufacturers to guarantee that its battery swapping solutions are compatible across various brands and models of passenger cars, thus providing users with a broader array of choices.
2. **Battery swapping products for light trucks, pickups, and micro-trucks in the logistics industry**
   For the logistics industry, CATL has introduced battery swapping solutions specifically designed for light trucks, pickups, and micro-trucks. These products are tailored to meet the demands of high-load operations and prolonged usage typical of logistics vehicles, with a strong emphasis on longevity and reliability.

   The long-life cell design features an operational lifespan of 8 years or 800,000 km while maintaining a State of Health (SOH) of at least 70%. This ensures that the battery continues to deliver optimal performance even under the rigorous conditions associated with logistics transportation. Additionally, a high-safety design is implemented to prevent thermal diffusion at elevated temperatures up to 35 °C, thereby ensuring safe operation for logistics vehicles.

   To address thermal diffusion concerns, it is recommended to utilize a heating trigger method followed by a 24-hour observation period after testing. In instances where a single battery experiences thermal runaway, an alarm signal can be activated promptly to mitigate risks associated with thermal diffusion and ensure that there is no fire or explosion risk externally.

6.1.1.2 Key Elements of Long Service Life and High Safety

1. **Long-life cell scheme**
   CATL's long-life cell initiative is founded on extensive research and development, as well as practical experience. The design objectives of 8 years, 800,000 km, and a SOH retention of 70% indicate that the battery can maintain 70% of its health status after eight years of usage.
   To achieve these ambitious goals, CATL has implemented advanced battery materials and manufacturing processes. For instance, the electrode structure has been optimized to enhance the charge-discharge cycle life of the battery; high-performance separators and electrolytes have been utilized to improve both stability and safety. Additionally, stringent quality control measures and detection methods are employed to ensure that each battery meets rigorous performance standards.
2. **High-safety design**
   Thermal diffusion represents a significant concern regarding battery safety. CATL effectively mitigates the risk of thermal diffusion through comprehensive system-level safety protection designs. By analyzing the energy distribution during thermal runaway in individual cells, key influencing factors include energy transfer to adjacent cells and energy dissipation via flue gas from thermal runaway events.
   CATL has implemented various strategies to minimize the risk associated with thermal diffusion. For example, when heat from an activated cell is directly transferred to neighboring cells due to large surface areas, specialized insulation materials and structural designs are employed to reduce both the speed and intensity of heat transfer. Furthermore, when an explosion-proof valve from an activated cell impacts the upper cover of a pack or when airflow from this valve affects internal components within the pack, potential thermal shock effects are alleviated by optimizing pack structures and incorporating protective layers.
   The BMS thermal runaway control strategy incorporates several functions, including RTC wake-up, AFE reverse wake-up, and air pressure sensor reverse wake-up. These features effectively address the thermal runaway inspection requirements across various scenarios.
   To mitigate the risk of combustion, an explosion-proof valve equipped with a rebound device is employed to limit the influx of oxygen. Additionally, insulation protection has been implemented on both the electrical connection output pole and connection terminal. A high-temperature-resistant insulating spray protective layer is applied to the surface of the high-voltage copper bar, while a PC flame-retardant insulation sheet covers the module. Furthermore, the upper cover of the battery pack consists of a stamped sheet metal design, collectively enhancing overall battery safety.

6.1.1.3 Mechanical Design and Battery Pack Design

1. **Mechanical design**
   CATL's mechanical design for battery swapping products is engineered to achieve a service life of 10 years and 1 million kilometers, thoroughly accounting

for mechanical stress and fatigue across various usage scenarios. The equivalent vibration test conducted is 3.4 times the duration specified by the new national standard, ensuring that the battery pack can endure diverse road conditions and transportation environments during prolonged use.

High-strength materials and advanced manufacturing processes are employed to enhance the structural integrity and stability of the battery pack. Additionally, both the installation method and fixation structure of the battery pack have been optimized to mitigate the effects of vibration and shock on its performance.

2. **Battery pack design**

   CATL's approach to battery pack design incorporates a long-life, high-reliability framework. By optimizing the battery management system, precise monitoring and maintenance of the battery are achieved, thereby enhancing both its service life and reliability.

   For instance, parameters such as battery power, voltage, and temperature are monitored in real-time. The charging and discharging strategies are adjusted based on the battery's state to prevent overcharging, over-discharging, and overheating. Concurrently, regular health checks and maintenance are conducted on the battery to identify and address potential issues promptly. For users, proper battery maintenance is of utmost importance. CATL emphasizes the significance of correct usage and upkeep by implementing a comprehensive battery full-life cycle monitoring system. For example, it is crucial to avoid over-discharging the battery since this can lead to irreversible physical damage. Additionally, minimizing frequent use of high-voltage fast charging is advisable because it generates excess heat; in the long term, this may accelerate battery aging and adversely affect both its overall service life and driving range. Even batteries of superior quality may sustain damage during their operational lifespan, resulting in users incurring additional costs for new batteries or facing safety hazards.

6.1.1.4 The Whole-pack LFP Battery Swapping Package Scheme

1. **Improving fast charging capability and service life**

   CATL has implemented a comprehensive whole-pack lithium iron phosphate (LFP) battery swapping package scheme aimed at enhancing fast charging capabilities and extending service life. LFP batteries are characterized by their high safety, long lifespan, and cost-effectiveness, making them well-suited for the battery swapping model.

   By optimizing both the structure and materials of the batteries, CATL has significantly improved the fast charging performance of LFP batteries. Additionally, an advanced battery management system is employed to maintain stable temperature and voltage during rapid charging processes, thereby minimizing potential damage to the batteries.

2. **Reducing costs**

   The whole-pack LFP battery swapping package scheme also contributes to cost reduction. The raw material costs associated with LFP batteries are relatively low, coupled with a mature production process that helps lower manufacturing expenses.

Moreover, the battery swapping model facilitates sharing and recycling of batteries, further decreasing user costs. By establishing a comprehensive system for battery recycling and reuse, CATL can effectively recycle retired batteries to extract valuable materials for use in new battery production—thereby promoting resource sustainability.

3. **Developing more competitive products**
   CATL is committed to continuous innovation and optimization in order to develop more competitive products within the realm of battery swapping technology. This commitment supports the promotion of this mode while addressing diverse user needs effectively.

   In addition to enhancing fast charging capabilities, extending service life, and reducing costs, CATL places a strong emphasis on the intelligence and convenience of its products. For instance, the company is developing an intelligent battery swapping system designed to facilitate automated battery exchanges and improve overall efficiency in this process. Simultaneously, by integrating Internet and Internet of Things technologies, CATL offers users a convenient battery swapping service along with a comprehensive management platform aimed at enriching the user experience.

   CATL is committed to continuous innovation and breakthroughs in the research, development, and application of battery swapping products. The company provides long-lasting, high-safety, and cost-effective solutions tailored for various usage scenarios. By establishing a full-life cycle monitoring system for batteries and strengthening operational maintenance management practices, CATL aims to extend battery lifespan while enhancing reliability and safety. Furthermore, the company strives to develop more competitive products that promote widespread adoption of the battery swapping model—contributing significantly to the sustainable development of new energy vehicles.

   The specification of a CATL LFP swappable battery pack is listed in Table 6.1.

In the domains of electric shipping and aviation, it has been reported that marine power lithium batteries and battery management systems (BMSs) are required to obtain type approval certificates from the China Classification Society (CCS) prior to installation on vessels. Data indicates that CCS has issued type approval certificates for lithium battery-related products to numerous manufacturers, including CATL, CALB, EVE Energy, Gotion High-tech, and Sunwoda. The certified products encompass battery cells, modules, BMSs, and complete battery packs.

As one of the earliest lithium battery manufacturers to enter the maritime sector, EVE Energy achieved CCS certification for its marine power batteries as early as 2016. Furthermore, in 2019, these batteries received explosion-proof certification from cc in Germany. Electric vessels such as "Princess Khutulun" and "Junlv" are equipped with power batteries supplied by EVE Energy.

CATL is also making significant strides in this field. In 2018, Fujian's first high-end electric inland cruise ship "Minjiang Star" embarked on its maiden voyage utilizing CATL's large-capacity lithium iron phosphate battery cells. Subsequently, in 2020 and 2021—as well as January 2022—CATL's lithium iron phosphate batteries were installed

Table 6.1 Specific scheme for LFP complete package swappable battery pack standard package.

Category	Project	Complete package swappable battery pack
System	Total energy (kWh@0.33C)	56
	Grouping method	1P114S
	Estimated weight (kg)	415
	System nominal voltage (V)	367.08
	Voltage range (V)	228–433.2
	Fast charging capability (30–80%SOC@25 °C)	25 min
	Pack weight energy density (Wh/kg)	135
	Design life	8 years, 800,000 km, 70% SOH
	DoD range	0–100%
	Thermal management solutions	Natural cooling + heating film

Source: CATL.

on vessels including "Deep Sea 01," "Zhong Shan Da Xue," and "Yangtze River Three Gorges No. 1." At the end of 2022, CATL established a wholly-owned subsidiary named CATL Electric Ship Technology Co., Ltd., focusing primarily on research, development, and application of technologies related to ship power battery systems.

The subsidiary of Gotion High-tech, Shanghai Gotion Wuyang Ship Technology Co., Ltd., secured its first batch of orders for three sets of ship power lithium battery systems in 2020. In July 2020, China witnessed the launch of its first hybrid leisure fishing boat powered by batteries. For this vessel, Gotion High-tech developed a highly integrated high-voltage control cabinet battery pack utilizing a lithium iron phosphate battery system. On October 13, 2020, the CCS officially awarded Gotion High-tech with the type approval certificate for "lithium iron phosphate battery and battery management system."

CALB's independently designed and developed zero-carbon ship energy storage power supply system is recognized as China's first ship battery system product to receive DNV certification from Det Norske Veritas. In 2023, China introduced an innovative electric catamaran, the 42-foot FAREAST 42C, which is also equipped with CALB's LFP battery products.

The research report titled "Market Prospect Research and Analysis Report on China's eVTOL Battery Market," released by GGII, indicates that the scale of the electric vertical takeoff and landing aircraft (eVTOL) battery market is expected to exceed one trillion yuan by 2040. As a core component of eVTOL technology, the performance and safety of batteries directly influence both the operational capabilities and market acceptance of eVTOL aircraft.

In comparison to electric vehicles, eVTOLs impose more stringent requirements regarding high specific energy, high power output, safety standards, rapid charging capabilities, and extended service life for their batteries. Furthermore, the specific power demands for batteries vary across different types of eVTOL during hovering and cruising phases. The level of specific energy (energy density) in these batteries ultimately determines the operational range of eVTOLs.

As they confront a burgeoning low-altitude flight market valued at trillions, numerous domestic power battery manufacturers have begun to venture into this emerging sector in pursuit of new opportunities presented by flying cars.

CATL, China's largest domestic power battery manufacturer, has long been strategizing its entry into the aviation sector. Notably, when it unveiled its condensed battery in April 2023, CATL expressed its ambition with the phrase "batteries taking to the sky." This innovative battery is designed not only for automotive applications but also for aviation use. At that time, Wu Kai—chief scientist at CATL—revealed that the company is actively engaged in collaborative development projects focused on civil electric manned aircraft.

Eve Energy has emerged as a leader in the mass production and delivery of power batteries for eVTOL. It was previously reported that the company has established a partnership with an internationally renowned eVTOL manufacturer and has commenced product deliveries. Wang Yu, the chairman of Eve Energy, stated during his appearance on the program "Dialogue" that "EVE Energy became involved in the field of aerial electrification five or six years ago."

At the end of 2023, Gotion High-tech and EHang Intelligent signed a strategic cooperation agreement. The two parties will collaboratively develop power battery cells, battery packs, energy storage systems, and charging infrastructure based on EHang Intelligent's eVTOL products while exploring new models for industrial collaborative development.

CALB is closely associated with XPeng Motors. The advanced 9-series high-nickel/silicon system battery developed for low-altitude travel represents a significant leap forward in lightweight design and safety performance while ensuring high power output and rapid charging capabilities.

Zenithen New Energy is focusing on developing power batteries specifically for the eVTOL market. Their technology achieves 20% state of charge (SOC) along with impressive discharge rates—12C for high-rate discharge and 5C for constant current discharge—while maintaining a high energy density of 320 Wh/kg. Currently, Zenithen New Energy is engaged in extensive collaboration with numerous leading domestic and international electric aircraft enterprises to pursue aviation airworthiness certification.

Lishen Battery has been actively exploring the aircraft electrification market for an extended period. Currently, the company has successfully implemented batteries with an energy density of 325 Wh/kg in eVTOL applications.

Firstly, regarding energy density, eVTOLs have a significantly higher demand for energy density compared to automobiles. Presently, the energy density of eVTOL batteries stands at 285 Wh/kg, which is 40% greater than that of ternary batteries used in passenger cars (200 Wh/kg) and twice that of lithium iron phosphate batteries utilized in passenger vehicles (140 Wh/kg). Nevertheless, even this performance index of 285 Wh/kg still falls short of meeting industry requirements.

Below are some prominent aviation battery suppliers worldwide:

AirEnergy: The company possesses extensive experience in developing battery systems for light aircraft. It specializes in designing battery enclosures as well as mechanical and electrical connections for battery packs and management systems tailored for aviation use. Each battery module is equipped with its own Battery Management System (BMS), allowing it to charge during flight. This capability enables the aircraft to approach landing areas with full power and optimal performance without needing to wait for charging between flights.

H55: Recently completed the first stage of Series C financing, securing $49 million USD. These funds will be allocated towards infrastructure investments, including establishing manufacturing facilities in Montreal, Quebec; advancing production capabilities in Switzerland and Canada towards semi-automated processes; and launching a subsidiary in Toulouse, France dedicated to promoting electric aviation technology.

INX: Established in 2020, this high-tech company specializes in the research, development, and production of solid-state lithium metal batteries. EHang has made strategic investments in this venture and plans to collaborate with INX to research, develop, and produce solid-state lithium metal storage batteries for EHang's aircraft products. The PreA+ financing round for INX is led by GL Ventures, with participation from Connected Intelligence Fund, Qianhai Fund, EHang itself, and existing investors such as Fenghe Capital. The funds raised will primarily be allocated towards the research and development, market expansion, production, and delivery of solid-state lithium metal battery technology.

Cuberg: Safran Power has entered into a cooperation agreement with Cuberg—a subsidiary of Northvolt—to jointly develop aviation energy storage systems intended for future all-electric and hybrid aircraft. Cuberg operates as a vertically integrated battery supplier that will manufacture and supply rechargeable aviation battery systems utilizing lithium metal technologies including cells, modules, and battery packs. Meanwhile, Safran Power will focus on designing global energy storage systems encompassing high-voltage protection components along with integration, support, certification, and maintenance services. Notably, the cells within Cuberg's battery system have been internally validated to achieve a specific energy output of 280 Wh/kg alongside an energy density of 320 Wh/L.

Northvolt: This Swedish manufacturer of lithium batteries acquired the American startup Cuberg in 2021 to leverage technology aimed at enhancing its batteries' lifespan. Since its inception, Northvolt has successfully raised over $3.6 billion USD in funding. Its objective is to compete against leading Asian battery manufacturers such as CATL and LG Chem while aiming for a 25% market share within Europe by the year 2030.

SolidEnergy Systems (SES): Based in Massachusetts, SolidEnergy Systems has developed an innovative new lithium metal anode battery specifically designed for eVTOLs as well as electric vehicles (EVs). This large-scale 100 Ah lithium metal battery boasts performance metrics exceeding 400 Wh/kg alongside an impressive energy density of up to 1,000 Wh/L.

As the application of batteries in the automotive sector continues to expand and mature, battery manufacturers are actively seeking new avenues for growth. The domains of electric shipping and electric aviation have emerged as promising areas of opportunity. Electric ships provide a more environmentally friendly and efficient transportation solution on water. With increasing certification of battery products by classification societies, an ever-growing number of battery enterprises are entering this market, thereby offering reliable power sources for maritime vessels.

In the realm of electric aviation, achieving higher energy density, enhanced safety, and rapid charging capabilities is paramount. Battery suppliers are consistently innovating and collaborating to meet the stringent requirements posed by eVTOL aircraft. The transition from automotive applications to those in maritime and aviation sectors signifies a new developmental phase for the battery industry, presenting both challenges and opportunities alike.

6.1.2 Charging Station Hardware Providers

In the battery swapping industry chain, hardware suppliers for battery swapping stations primarily occupy an upstream position and are responsible for providing essential hardware equipment necessary for these stations. These suppliers typically design and manufacture a variety of hardware components, including quick-change systems, charging equipment, control systems, and more.

The battery swapping system is mainly composed of a battery swapping platform, a charging platform, and a control module for both charging and battery swapping. Passenger vehicle battery swapping stations predominantly feature automated equipment (accounting for approximately 50% of the total value), such as RGV carts, unlocking and locking platforms, hydraulic lifts, etc.; charging and discharging equipment (representing about 20% of the total value), which includes charging cabinets and modules; containers (making up around 15% of the total value); along with various station control systems encompassing temperature regulation, fire protection measures, and monitoring systems.

Currently, domestic suppliers of battery swapping equipment include Weida Group, BOZHON Precision Industry, Harmontronics, and CSG Smart Science & Technology Co., Ltd. Notably, Weida Group has established a close partnership with NIO by jointly founding a subsidiary named Swap that exclusively supplies battery swapping equipment for NIO's second-generation stations. Meanwhile, BOZHON Precision Industry, Harmontronics, and CSG Smart Science & Technology Co., Ltd possess substantial technical expertise in the field of battery swapping equipment. Their products have undergone several generations of enhancements to provide solutions not only to NIO but also to traditional automotive manufacturers such as Dongfeng Motor Corporation (DFM), BAIC Group (Beijing Automotive Industry Holding Co.), and Geely Automobile Holdings Limited.

The core competitiveness of battery swapping equipment enterprises is rooted in their comprehensive strengths in both hardware and software. Currently, suppliers of battery swapping equipment primarily offer complete battery swapping stations. Key manufacturers in this sector include companies such as Harmontronics, BOZHON, Weida Group, and CSG Smart Science & Technology Co., Ltd.

In terms of hardware, battery swapping equipment is utilized across various outdoor scenarios. This application imposes stringent requirements on product design, safety, and reliability; consequently, the verification cycle tends to be relatively lengthy. Furthermore, battery swapping stations incorporate multiple technologies—including precision machinery, electronic control systems, and industrial Internet—which present certain technical barriers to entry.

Once successfully established within the battery swapping market, competition among equipment enterprises increasingly emphasizes soft power elements such as capital investment, mass production capabilities, and delivery efficiency, as well as strategic partnerships with major downstream customers. For instance, Harmontronics has entered into a five-year long-term strategic cooperation agreement with GCL Energy Technology Co., Ltd., while Weida Group maintains a strong partnership with NIO. By leveraging deep collaborations with key downstream clients alongside robust batch delivery capabilities, it is anticipated that battery swapping equipment enterprises will experience significant performance enhancements. Some well-known examples have been summarized in Table 6.2.

Table 6.2 Downstream channels and existing battery swap station products.

Downstream channel	OEM partners	Existing battery swapping station products
Harmontronics	Dongfeng, Liuqi, Chery, BAIC, FAW, etc.	Includes subcontracting battery swapping and complete battery swapping station products. Battery packs ranging from 7 to 48 can be combined freely, suitable for battery swapping of passenger vehicles and commercial vehicles.
BOZHON	NIO, BAIC, Geely	Can complete battery swapping within 90 seconds. The maximum daily service capacity is 400 times. A single station contains 26 battery packs and can realize fully automatic unattended battery swapping.
Weida Group	NIO	NIO's second-generation battery swapping station has 13 battery compartments. The battery swapping time is 4.5 minutes. The maximum daily service capacity is 312 vehicles. It adopts chassis-type battery swapping.
CSG	NIO, Geely, etc.	Adopts chassis battery swapping technology. The battery swapping process only takes one minute. Adopts a highly compatible modular architecture of the battery swapping platform. The battery swapping station can be adapted to multiple vehicle models.

6.1.2.1 Harmontronics

Harmontronics, established in 2007, is a professional provider of comprehensive intelligent manufacturing solutions. The company went public on the STAR Market on July 22, 2019, becoming one of the first enterprises listed on this platform.

Focusing on the electrification and intelligent development of automobiles, Harmontronics offers an extensive range of solutions and services that include automation equipment, core components, and industrial software tailored for industries such as automotive and new energy. The company has mastered key technologies in intelligent manufacturing equipment and systems. Currently, it holds 588 core intellectual property rights, including 27 invention patents. Harmontronics Intelligent occupies a leading position in the domestic market share within the niche sector of automotive intelligent manufacturing equipment and is among the few suppliers capable of participating in international projects related to intelligent manufacturing equipment; its business operations extend across more than 20 countries.

Specializing in intelligent manufacturing for new energy vehicle charging and swapping systems, Harmontronics provides products and solutions such as passenger vehicle swap stations, commercial vehicle swap stations, powertrain battery boxes, and swap station management platforms—alongside charging products that include charging machines, charging piles, and mobile energy vehicles. These offerings deliver comprehensive services across multiple dimensions—including cloud platforms, station-side control mechanisms, core component development and production processes—as well as operational services and maintenance management. This ensures customers receive safe, efficient, and reliable charging and swapping products along with exceptional service quality.

These charging and battery swapping products are designed for a variety of applications, including commercial complexes, large residential areas, extensive parking facilities, service zones, ports, steel manufacturing plants, coal mines, urban construction waste transportation systems, public-rail transfer logistics, and highway freight transport. In response to the rapid advancement of new energy technologies, Harmontronics has refined its corporate strategy. The company is now concentrating on key business channels such as automotive intelligence and electrification while establishing three primary business segments: automotive intelligent equipment, battery intelligent equipment, and charging and battery swapping solutions.

Leveraging its extensive experience in automotive equipment manufacturing accumulated over the years, Harmontronics initiated its foray into the charging and battery swapping sector in 2020 and commenced commercial operations in 2021. Its product offerings encompass three main categories: manufacturing of battery swapping station equipment; production of core components; and development of terminal operation system platforms. These products cater to a wide range of application scenarios involving passenger vehicles, light trucks, heavy-duty trucks, mining vehicles, drones, and maritime vessels.

Harmontronics closely monitors the development pace of the new energy vehicle industry and actively adjusts its strategic focus. In response to the rapid advancements in this sector, Harmontronics made a significant adjustment to its “1+N” strategic model in 2021: prioritizing core business channels such as automotive intelligence and electrification while expanding its original automotive segment to encompass new energy three-electric systems and thermal management system businesses. The former automotive electronics business unit (BU) has been rebranded as the automotive BU; concurrently, the charging and battery swapping operations from the previous new energy segment have been established as an independent charging and battery swapping BU, aimed at vigorously expanding into the automotive battery swapping market. Additionally, what was once known as the new energy BU is now referred to as the battery equipment BU. At this juncture, noncore businesses have been phased out to further concentrate on primary operations.

The battery swapping station is primarily designed to achieve functions such as accurate identification and positioning of vehicles, precise control of mechanical transmission systems, stable and reliable mechanical–electrical connections, equipment status detection and monitoring, along with cloud interaction capabilities. Through years of accumulated expertise in automotive electronic equipment, Harmontronics possesses substantial reserves in foundational general technologies related to machine vision, high-speed positioning technology, ultra-high-speed precision curved conjugate cam technology, energy feedback towing test technology, high-speed pressure and displacement detection technology, as well as an industrial Internet research and development platform. This robust technological foundation ensures that Harmontronics can swiftly enter this emerging field while establishing a first-mover advantage.

6.1.2.1.1 Product Examples

An example of Harmontronics U1 series passenger vehicle battery swapping station is illustrated in Figure 6.3.

Features:

- have corresponding mature solutions for alpine, coastal, and high-altitude areas to meet customer market demands and achieve full regional coverage;
- can realize compatible battery swapping for mainstream A-class to A0-class models in the market, improving operational revenue;
- unattended, with remote monitoring upgrade. Build a service system and basic information management to realize intelligent after-sales service.

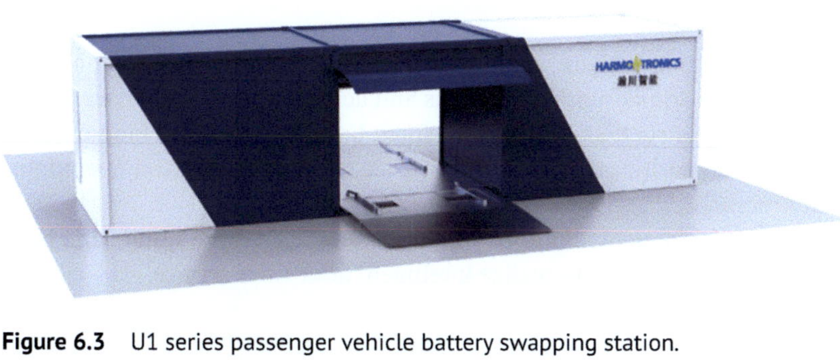

Figure 6.3 U1 series passenger vehicle battery swapping station.

Figure 6.4 U2 series light truck battery swapping station.

The Harmontronics U2 series light truck battery swapping station is shown in Figure 6.4.

Features:

- have corresponding mature solutions for alpine, coastal, and high-altitude areas to meet customer market demands and achieve full regional coverage;
- adapt to vehicle postures and be compatible with multiple wheelbases;
- unattended, with remote monitoring upgrade. Build a service system and basic information management to realize intelligent after-sales service.

Harmontronics has also developed different battery swapping stations specific to heavy trucks, as illustrated in Figures 6.5–6.7.

Features:

- intelligent battery swapping robot adapts to vehicle postures and has intelligent torque control technology for unlocking and locking;
- have corresponding mature solutions for alpine, coastal, and high-altitude areas to meet customer market demands and achieve full regional coverage;
- intelligent whole-station control system and maintenance warning technology to improve operational stability;
- unattended, with remote monitoring upgrade. Build a service system and basic information management to realize intelligent after-sales service.

Features:

- have corresponding mature solutions for alpine, coastal, and high-altitude areas to meet customer market demands and achieve full regional coverage;
- have technologies such as lifting point force balance detection, long-distance full-closed-loop positioning, and maintenance warning to improve operational stability;
- unattended, with remote monitoring upgrade. Build a service system and basic information management to realize intelligent after-sales service.

Figure 6.5 U3 series heavy truck battery swapping station.

Figure 6.6 T1 series heavy truck battery swapping station.

Figure 6.7 T2 series heavy truck battery swapping station.

Features:

- have corresponding mature solutions for alpine, coastal, and high-altitude areas to meet customer market demands and achieve full regional coverage;
- convenient transportation and efficient station restoration;
- realize separation of battery charging and swapping, safe, and reliable.

Specific battery swapping stations for special vehicles, e.g., mining trucks, have been developed, as illustrated in Figure 6.8.

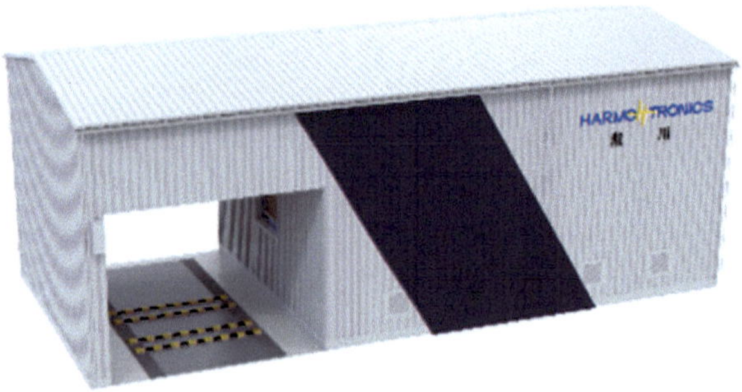

Figure 6.8 T4 series mining truck battery swapping station.

Table 6.3 Locking mechanism type–application reference.

Type	Key features	Typical applications
Steel Ball Type	Steel-ball structure enhances stability	Battery swapping robot
Bolt Type	High durability and stable locking force	Lifting tool
Rotary Down-Pressure	New anti-jamming technology for higher safety	Unlock-lock RGV
Pin Type	Locking solution: pretension + brake self-locking	On-vehicle lifting frame

Features:

- frame structure design, stable and reliable;
- modular design of core components to ensure rapid delivery;
- unattended, with remote monitoring upgrade. Build a service system and basic information management to realize intelligent after-sales service.

Features:

- can be compatible with mainstream market products;
- high-reliability locking attachment. Extremely stable battery swapping controller, effectively waterproof and collision-proof;
- adopt new foolproof technology, high safety. Have good durability and stable locking force, long service life, and easy maintenance.

There're four major locking structures that have been widely used, as listed in Table 6.3.

6.1.2.1.2 Core Components

Based on product performance, we will develop standardized common components and mass-produce them to achieve cost reduction.

A comprehensive monitoring system for the operation and management of the entire battery swapping station. Realize charging and battery swapping strategy management, vehicle battery swapping control management, battery charging management, equipment management, station wireless interconnection, cloud platform communication, etc. At the same time, support real-time monitoring and reporting of faults and fire safety in the station, support rapid identification and handling of events, improve the work efficiency of on-duty personnel in the station, and ensure the safe and efficient operation of the entire station.

6.1.2.1.3 Cloud Platform Management System

This is a comprehensive management platform designed for the operational terminals of charging and battery swapping stations. It enables unified management of assets, including personnel, vehicles, stations, equipment, and batteries. The platform supports operational strategy management as well as order processing capabilities. Furthermore, it provides graphical statistical monitoring of production data while facilitating connections with third-party data sources. This solution assists enterprises in addressing challenges such as dispersed battery swapping stations, disorganized management practices, prolonged problem resolution cycles, and unclear operational outcomes—ultimately reducing operating costs while enhancing overall operational efficiency.

The charger can be categorized into vertical or horizontal, as illustrated in Figures 6.9 and 6.10.

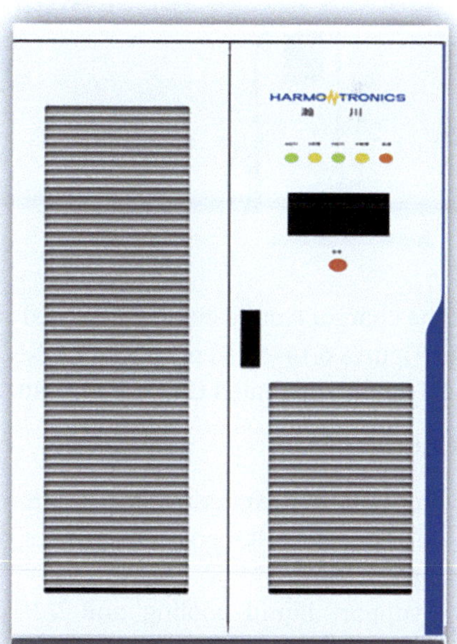

Figure 6.9 240/300/360 kW vertical charger.

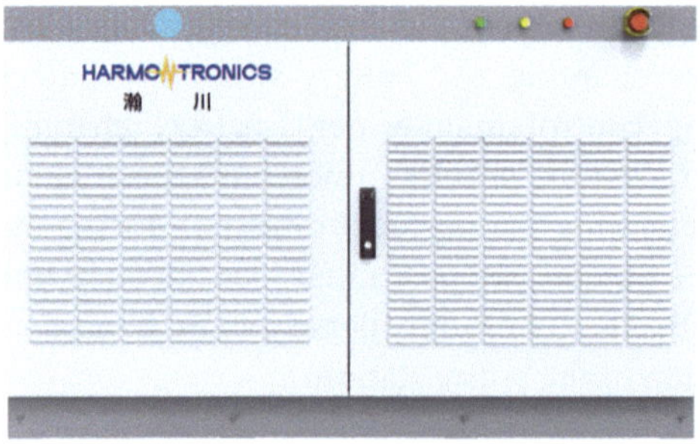

Figure 6.10 240/300/360 kW horizontal charger.

Figure 6.11 H3 series 120/160 kW integrated charging pile.

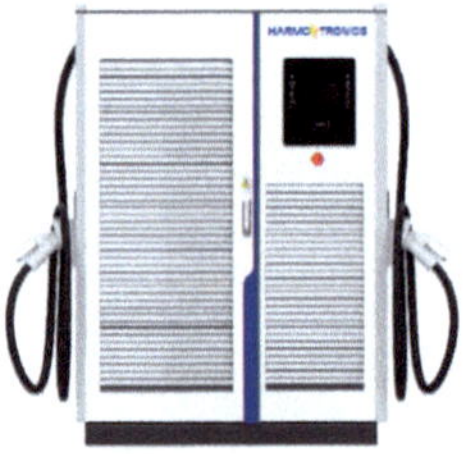

Figure 6.12 H3 series 240/320 kW integrated charging pile.

The charger can be either integrated or spilt, as can be seen in Figures 6.11–6.13 and Figures 6.14–6.16 correspondingly.

H1 Series Integrated Charger for Battery Swapping Station

Features:

- optimized heat dissipation design, stable and reliable;
- support simultaneous charging of batteries and vehicles, with a high resource reuse rate;
- support liquid cooling and V2G (bidirectional charging and discharging control);
- scenario: Commercial vehicle battery swapping station.

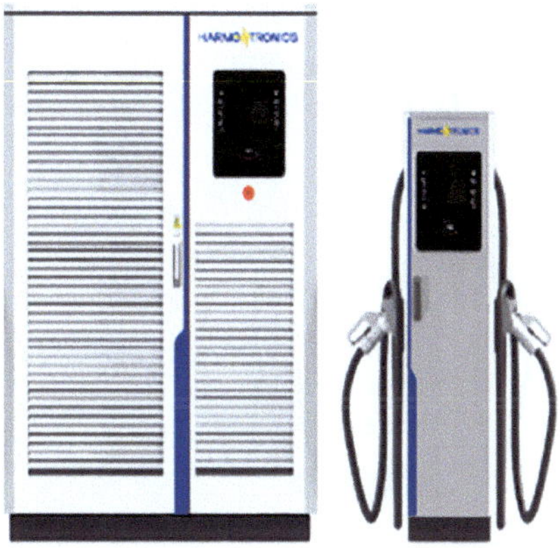

Figure 6.13 H3 series 300/360kW integrated charging pile.

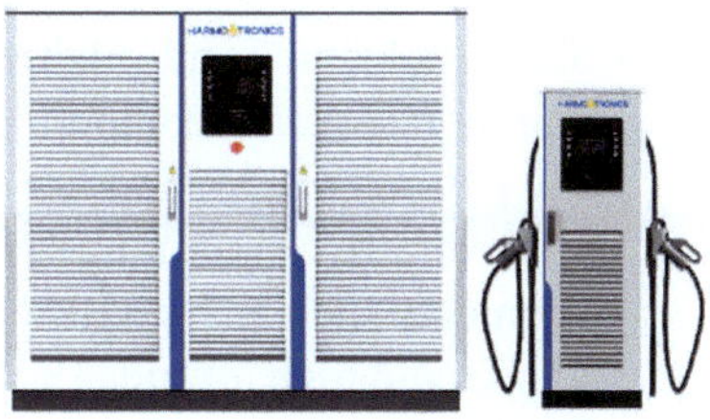

Figure 6.14 H4 series 360/480 kW split-type group charging pile.

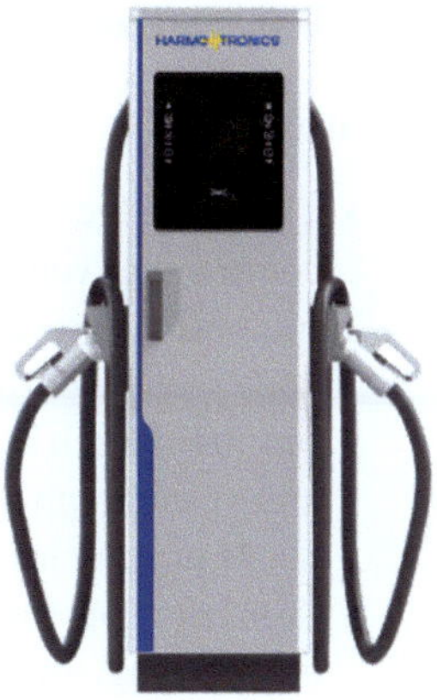

Figure 6.15 H4 series 600/720/960 kW split-type group charging pile.

Features:

- The intelligent and efficient DC charging pile is mainly suitable for high-power charging scenarios. The module adopts a full potting process and has strong environmental adaptability. It has an ultra-wide voltage output range of 50–1000 V, meeting the charging needs of various vehicle models. It is more reliable, more complete, and more durable.
- Scenario: Residential areas, commercial places, enterprises and institutions, scenic spots, industrial parks, and parking lots.

Figure 6.16 H5 series split-type group charging pile.

Features:

- The intelligent and highly reliable DC charging pile is customized for commercial vehicle charging. The charging module adopts a high-protection full potting process and has strong environmental adaptability. It has an ultra-wide voltage output range of 50--1000 V and can be widely used in harsh environments such as high dust (mines, steel mills, etc.), strong corrosion (coastal areas), and high altitude (Sichuan-Tibet line). It mainly solves the fast charging needs of vehicles with battery capacities of 282 kWh and below.
- Scenario: Port transportation, steel enterprises, mine transportation, municipal engineering, and short-distance feeder transportation

Features:

- The intelligent and highly reliable DC charging pile is customized for commercial vehicle charging. The charging module adopts a high-protection full potting process and has strong environmental adaptability. It has an ultra-wide voltage output range of 50–1000 V and can be widely used in harsh environments such as high dust (mines, steel mills, etc.), strong corrosion (coastal areas), and high altitude (Sichuan-Tibet line). It mainly solves the fast charging needs of vehicles with battery capacities of 282–422 kWh.
- Scenario: Port transportation, steel enterprises, mine transportation, municipal engineering, and short-distance feeder transportation.

Features:

- The split-type group charging pile is composed of a main power cabinet and 3 to 12 charging terminals. The product adopts a modular design concept. With the full-matrix power distribution module with full potting and high protection as the core, it uses flexible charging technology to realize efficient power sharing and on-demand distribution, which can meet the fast charging needs of electric vehicles of different models and powers. This product uses a multiple thermal management system to ensure stable operation of the charging pile, faster charging speed, and greater safety. Super strong adaptability, compatible with charging of various vehicle models such as buses, passenger vehicles, and logistics vehicles, with better compatibility.

- Scenario: Shopping malls/hotels, enterprises and institutions, parks/scenic spots, industrial parks, parking lots, port transportation, steel enterprises, mine transportation, municipal engineering, and short-distance feeder transportation.

Features:

- Adopt a high-definition color liquid crystal display screen, easy to operate, friendly interface. The charging experience is more concise and efficient. With an IP54 protection level. The high and low voltages are designed with isolated partitions. Integrated with multiple charging protection technologies, it has overvoltage and under-voltage protection, leakage current protection, high and low temperature protection, lightning protection, etc. The control circuit and power circuit are completely isolated to enhance anti-interference ability and fully ensure the stable operation of the circuit.
- Scenario: Shopping malls/hotels, enterprises and institutions, parks/scenic spots, industrial parks, parking lots, port transportation, steel enterprises, mine transportation, municipal engineering, and short-distance feeder transportation.

6.1.2.2 BOZHON Precision Industry Technology Co., Ltd.

BOZHON Precision Industry, established in 2006, specializes in the research and development, design, production, sales, and technical services of automated equipment. The company focuses on automated flexible production lines, key components for automation, and tooling fixtures. Its product offerings primarily span the consumer electronics, automotive, and new energy sectors.

In the new energy industry, BOZHON's core products include automotive parts production equipment as well as standard machinery for lithium battery manufacturing—such as liquid injection machines—and automated production lines for modules/PACKs. Additionally, the company develops intelligent charging stations and battery swapping solutions.

With a strong commitment to intelligent manufacturing equipment, BOZHON Precision Industry offers a diverse range of products that encompass automated systems (lines), fixture solutions, and essential component products. The company is recognized as a leader in the consumer electronics sector while continuously pursuing horizontal expansion and vertical integration of its product line.

Leveraging its accumulated expertise and advantages in key technologies within the realm of automation equipment, BOZHON Precision Industry has successfully diversified its business into emerging fields such as new energy and semiconductors.

In 2018, BOZHON Precision Industry ventured into the realm of charging and battery swapping stations, providing fully automated solutions for NIO, which was the pioneer in proposing the battery swapping model. By 2021, the company had successively launched a range of products, including standard and customized battery swapping stations, heavy truck battery swapping stations, as well as automatic charging and battery swapping stations designed specifically for heavy trucks. Currently, these offerings have evolved to their seventh generation. The daily maximum service capacity of passenger vehicle battery swapping stations can reach between 400 and 600 swaps.

At present, BOZHON has established partnerships with notable clients such as BAIC Blue Valley, Geely Automobile, CATL, Dongfeng Motor Corporation, and GCL Energy Technology Co., Ltd. As of Q1 2022, the order volume for charging and battery swapping station equipment on hand amounted to approximately 41.6317 million yuan.

In January 2022, the company announced a fixed increase plan aimed at raising 1.7 billion yuan; out of this amount, 746 million yuan is allocated for expanding and constructing automated equipment within the new energy sector. Upon completion of production enhancements under this initiative, an additional capacity will be introduced, comprising 400 sets of passenger vehicle battery swapping stations and 100 sets designated for commercial vehicles.

6.1.2.3 Weida Group

Weida Group is a prominent enterprise in the field of power tool accessories. In 2016, the company initiated research and development on battery swapping technology through its acquisition of Suzhou Demai Ke, thereby entering the new energy battery swapping sector. Currently, Weida Group's primary business encompasses power tools, high-end intelligent equipment manufacturing, and the new energy industry. Its product offerings include power tool accessories, battery packs, battery swapping stations, charging piles, powder metallurgy components, machine tools, and more.

Founded in 1998 with a focus on producing drill chucks and their associated accessories, Weida Group has progressively evolved into a leading player in the power tool accessory market. Following its listing in 2004, the company began to pursue diversified industrial development by acquiring Shanghai Baichi, Jinan No.1 Machine Tool Company, and Suzhou Demai Ke to expand into intelligent manufacturing and new energy sectors.

As the new energy vehicle industry experiences significant growth, Weida Group commenced its exploration of battery swapping technology research and development in 2016. Subsequently, in August 2017, it established Swap Intelligent Equipment Co., Ltd. in collaboration with NIO.

Demai Ke holds 44% of the shares, while NIO Energy possesses 43.5%. The primary products offered include battery swapping stations, intelligent modules, charging, and battery swapping cabinets, as well as DC and AC charging piles. As the exclusive supplier of NIO's battery swapping stations, it has collaboratively developed and launched both the first and second generations of these stations with NIO.

In January 2021, Sowepower entered into a strategic cooperation agreement with Zhejiang Jianeng to jointly expand the market for automated battery swapping services in new energy vehicles. Furthermore, Sowepower is actively broadening its downstream customer base and has successfully engaged in business collaborations for battery swapping stations with State Grid, China Southern Power Grid, Zhejiang Rongxiang, and the Chinese Academy of Sciences.

NIO stands out as a leading enterprise in domestic C-end battery swapping solutions. The market share within this sector continues to grow steadily. By 2025, it plans to establish a network of 4,000 battery swapping stations globally.

6.1.2.4 CSG Smart Science & Technology Co., Ltd

CSG Smart Science & Technology Co., Ltd. was established in 2002 and has emerged as one of the leading domestic providers of industrial intelligent solutions. The company's primary business areas encompass industrial automation and power automation, with a strong emphasis on product research, development, and application across various sectors including industrial robotics, intelligent logistics, power systems, and new energy.

Currently, the battery swapping products are primarily managed by its wholly-owned subsidiary, Shanghai Yongqian Electromechanical Co., Ltd. Yongqian Electromechanical's core operations include industrial robots, intelligent production lines, and smart factory information systems. At present, the company's battery swapping station offerings cater to a diverse range of vehicles such as passenger cars, heavy trucks, logistics vehicles, and sanitation vehicles.

As an early entrant into the battery swapping sector in China, CSG possesses significant first-mover advantages in research and development capabilities, technological foresight, talent acquisition and retention, as well as product iteration and enhancement. The independently developed new energy vehicle battery swapping stations have evolved from the first-generation manual models to fifth-generation unattended intelligent stations.

At this time, key customers for CSG's battery swapping products include prominent companies such as Geely and NIO.

6.1.3 Battery Management System and Related Software Technology Suppliers

Battery management system, commonly referred to as BMS, is a fundamental component of electric vehicle battery packs. Its primary function is to monitor and control the operational state of battery packs, ensuring their safe and efficient performance. The BMS serves as a crucial link between the batteries and the entire vehicle system.

In battery swapping mode, the BMS has dual responsibilities. Firstly, it monitors and manages various parameters critical to battery health, including voltage, current, temperature, SOC, and SOH. This monitoring ensures that batteries operate within designated safety limits. Secondly, it facilitates communication with the control system of the battery swapping station by transmitting essential battery information. This enables accurate identification of battery status at the swapping station and supports effective allocation and charging management.

During the process of battery swapping, it is imperative for the BMS to guarantee secure and reliable connections between batteries while preventing hazardous situations such as arcing or short circuits. Additionally, through real-time monitoring and analysis of battery conditions, the BMS can promptly identify potential issues such as overcharging, over-discharging, or overheating. It then implements appropriate protective measures to prevent damage to batteries while extending their lifespan.

Furthermore, the BMS achieves balanced management across all batteries in a pack to ensure uniform power distribution among individual cells. This balance enhances overall performance and extends driving range capabilities for electric vehicles.

Taking NIO as an example, the battery real-time monitoring system (BRMS) within NIO's BMS is primarily tasked with monitoring and predicting safety-related conditions. Through collaboration with the after-sales team and battery swapping operations, any potential issues with the battery pack can be addressed proactively before abnormalities arise. This allows engineers to promptly identify the root cause of any irregularities and develop appropriate improvement plans. Additionally, NIO's BMS encompasses functionalities such as battery equalization and capacity evaluation to ensure both safety and consistency during the battery swapping process.

With the rapid advancement of the electric vehicle market and the increasing adoption of battery swapping technology, there is a notable upward trend in both BMS and related software technology markets. In recent years, production and sales of new energy vehicles in China have surged significantly, leading to heightened demand for BMS solutions. According to relevant data, in 2022, approximately 7.0582 million automotive battery management systems were demanded in China—a year-on-year increase of 99.1%. The market size reached approximately 19.307 billion yuan, reflecting a year-on-year growth rate of 87.3%.

Given that the battery swapping model imposes stricter requirements on battery consistency and reliability, this necessitates ongoing upgrades and innovations in BMS technologies as well as associated software solutions. For instance, advancements are needed in state estimation algorithms to enhance accuracy regarding SOC and SOH. Furthermore, improvements in communication technology are essential for achieving faster and more stable data transmission capabilities that meet real-time demands during the battery swapping process. Lastly, enhanced fault diagnosis mechanisms along with early warning systems are critical for ensuring safety throughout all stages of operation within this context.

6.1.3.1 Requirements for BMS in the Battery Swapping Mode

1. **Real-time monitoring and fault diagnosis**
   - Real-time monitoring: BMS needs to monitor parameters such as battery voltage, current, and temperature in real time to timely discover potential abnormal situations.
   - Fault diagnosis: When an abnormality is detected, BMS should be able to quickly diagnose the cause of the fault and take corresponding measures to prevent the situation from worsening.
2. **Battery consistency management**
   - Battery equalization: During the battery swapping process, capacity differences between different batteries may lead to performance degradation. BMS needs to have an efficient battery equalization function to ensure the consistency of all batteries.
   - Capacity assessment: Accurately assess the remaining capacity of each battery pack to reasonably arrange the usage order of batteries.
3. **Safety protection**
   - Overcharge/overdischarge protection: BMS needs to have the function of preventing battery overcharge or overdischarge to ensure that the battery will not be damaged due to extreme conditions.
   - Thermal management: Adjust the battery temperature through cooling or heating systems to avoid damage to the battery caused by excessive or too low temperatures.
4. **Data communication and interaction**
   - Remote monitoring: BMS should support remote monitoring function so that the battery swapping station can obtain battery state information in real time.
   - Data upload: Be able to upload battery operation data to the cloud to provide a basis for subsequent analysis and maintenance.
5. **Reliability and durability**
   - High reliability: BMS needs to have high reliability to ensure stable operation in various harsh environments.
   - Long service life: Considering the frequent use of the battery swapping mode, the durability of BMS is also one of the important considerations.

   BMS technology development trend.

   With the progress of technology, BMS is also constantly developing and improving. In the future, BMS technology will develop in the following directions:

   - Intelligence: Integrate artificial intelligence technology to improve the intelligent analysis ability and fault prediction accuracy of BMS.
   - Modularization: Adopt a modular design to facilitate the upgrade and maintenance of BMS.
   - Wireless communication: Use wireless communication technology to reduce cable connections and improve the flexibility and scalability of the system.

6.1.3.2 Benefits of BMS in Combined Battery Swapping Mode

1. Battery state estimation can be more accurate. During the process of receiving charging for battery blocks in the battery swapping station, the battery swapping station has the ability to use stronger computing power and more complex state estimation algorithms to correct the battery state data.
2. Battery equalization can be more efficient. The battery equalization module in the traditional battery management system is limited by the equalization power and is often very inefficient. However, in the battery swapping station, there are more abundant conditions for battery equalization.

As an important link connecting battery packs, the entire vehicle system, and the motor, BMS is the core technology of power battery packs and also the most important indicator for measuring the performance of a pure electric vehicle.

The complexity of the BMS system itself determines that it has a strong technical barrier and requires technical research and development time. Its cost accounts for about 20% of the total cost of the battery pack.

Chinese BMS enterprises can generally be categorized into several types: self-operation by battery manufacturers, self-operation by vehicle manufacturers, and third-party operations.

Self-operation by battery manufacturers: Leading domestic power battery companies such as CATL and Gotion High-tech exemplify this category. These enterprises possess comprehensive core technologies, granting them a significant competitive edge in the market. The BMS developed by these companies is typically tailored to their own batteries, facilitating an optimized integration of battery performance with management systems. This results in enhanced performance and stability. For instance, CATL not only holds a prominent position in the research and development as well as production of batteries but also allocates substantial resources towards innovation in BMS technology, thereby providing reliable management solutions for its battery products.

Self-operation by vehicle manufacturers: This category is represented by companies such as BYD, BAIC New Energy, and Zhongtong Bus. In addition to possessing core technological expertise, these manufacturers enjoy cost advantages over others due to their ability to closely integrate BMS development with vehicle design and production processes. This integration optimizes the overall performance of the vehicles they produce. For example, BYD has accumulated extensive technical knowledge in the field of new energy vehicles; its independently developed BMS is well-suited for its electric vehicles and demonstrates excellent capabilities regarding vehicle power performance, driving range, and battery safety management.

6.1.3.3 Third-party Suppliers

Professional BMS suppliers: Notable enterprises such as Dongguan Powerway, Huizhou EVE Energy, and Shenzhen Kelit Technology are prominent players in the BMS sector. These companies emphasize research and development, alongside the production of BMS and associated software technologies. They typically possess robust technical expertise and innovative capabilities, enabling them to offer customized solutions tailored to various battery and vehicle manufacturers. For instance, Huizhou EVE Energy boasts extensive experience and a highly skilled technical team in the BMS domain, with its products enjoying significant recognition within the market.

Automotive parts suppliers: Major automotive parts corporations like Bosch and Delphi also hold a considerable share of the BMS market. Leveraging their research and development strengths in automotive components, along with long-standing collaborative relationships with vehicle manufacturers, these companies engage actively in the R&D and production of BMS solutions. Such enterprises generally have established production processes coupled with comprehensive quality control systems that ensure the reliability and excellence of their BMS products.

6.1.3.4 Global BMS Battery Swapping Suppliers' Information

In addition to the aforementioned Chinese suppliers, there exists a range of well-known global providers specializing in BMS battery swapping technology.

LG Chem: A prominent South Korean company recognized for its cutting-edge battery technology. Its BMS solutions are extensively utilized in electric vehicles globally. LG Chem's BMS is distinguished by its high accuracy in state estimation and exceptional reliability, effectively managing the battery pack to ensure both safety and performance in electric vehicles.

Panasonic: A leading Japanese electronics corporation with a rich history in battery technology. Panasonic's BMS is highly integrated, offering efficient battery management across various types of electric vehicles. The company emphasizes continuous innovation to enhance both the performance and lifespan of its batteries.

Tesla: While Tesla is primarily renowned as an electric vehicle manufacturer, it also develops its proprietary BMS. Tesla's BMS is meticulously optimized for its specific battery packs and vehicles, delivering outstanding performance and range. It employs advanced algorithms alongside real-time monitoring to guarantee the safety and longevity of the battery system.

6.2 Midstream Battery Swapping Operators and Electric Transportation Equipment Manufacturers

6.2.1 Battery Swapping Service Providers

6.2.1.1 Electric Two- and Three-wheelers

Electric two- and three-wheeler battery swapping operators are specialized enterprises that provide battery swapping services for electric two-wheelers (E2W), such as electric bicycles and motorcycles, as well as electric three-wheelers (E3W). These operators establish comprehensive battery swapping networks to offer users convenient services, effectively addressing range anxiety issues associated with the use of electric two- and three-wheel vehicles.

China Tower: In 2019, China Tower established a wholly-owned subsidiary, Tower Energy Co., Ltd., dedicated to the battery swapping business. Tower Energy has introduced integrated battery swapping solutions including “intelligent battery swapping cabinets, intelligent lithium batteries, and an intelligent operation platform for battery swapping applications.” The company has constructed over 340 “rider homes” nationwide, providing one-stop services encompassing rest areas, battery swaps, rentals, and repairs specifically tailored for food delivery riders and other outdoor workers.

Zhizu Tech: Zhizu Tech primarily focuses on safe and intelligent charging and swapping services for two-wheeled electric vehicles. Additionally, it offers rental and sales platform services for these vehicles along with IoT cloud management platform services and online big data analytics. Zhizu Tech’s cabinets are strategically located around food delivery stations, residential communities, office buildings, shopping malls, supermarkets, convenience stores, and farmers’ markets to deliver essential “endurance” support for food delivery riders.

This Lithium Battery Swapping service is a platform brand developed by Hangzhou Yugu Technology Co., Ltd., dedicated to providing safe and convenient battery swapping solutions for owners of two-wheeled electric vehicles, delivery riders, and intra-city delivery personnel nationwide.

Battery swapping platforms mainly targeting the C-end market are as follows:

Dudu Huandian: This is a battery swapping brand under Shenzhen Dudu Internet of Things Technology Co., Ltd. By integrating electric bicycles, batteries, and battery swapping cabinets through the Internet of Things (IoT), Dudu Huandian has established China’s first community-oriented safety battery swapping IoT platform for electric bicycles, offering both software and hardware solutions. It is reported that the monthly subscription prices offered by Dudu Huandian range from 20 to 50 yuan, with charges based on actual electricity consumption. This pricing model aligns well with the short-distance travel needs of community residents engaged in activities such as commuting to work, picking up and dropping off children, grocery shopping, and sending or receiving packages. In terms of compatibility with various models, Dudu Huandian’s lithium batteries are compatible with over 90% of electric vehicle models available in the market.

At the outset of the development of the two-wheel electric vehicle battery swapping industry, with the exception of Dudu Huandian, most battery swapping platforms primarily targeted rider groups by offering instant delivery services. As the penetration rate of battery swapping services increases and the user base expands, data indicates that by 2025, it is anticipated that the penetration rate in B-end markets—such as food delivery and express logistics—will exceed 50%, thereby providing battery swapping services to approximately 8– 10 million B-end customers. In contrast, in C-end markets, this penetration rate is expected to surpass 10%, catering to more than 30 million C-end users.

Recognizing the potential within the C-end battery swapping market, many platforms have begun to extend their services accordingly. Notable among these is Hellobike Battery Swapping: a "two-wheel battery swapping service" platform collaboratively launched by Hello Inc., CATL, and Ant Financial. For B-end users, Hellobike Battery Swapping offers a monthly package allowing unlimited swaps for around 300 yuan; meanwhile, for C-end users, new monthly packages have been introduced featuring options such as ten swaps for just 49 yuan. Regarding compatible models, Hellobike Battery Swapping supports compatibility with most mainstream brand models.

Aihuanhuan: Aihuanshuan Energy Co., Ltd. is an intelligent battery swapping service operator that leverages the globally leading intelligent energy platform, Gogoro Network. The company integrates advanced research and development in intelligent batteries, energy storage systems, battery swapping stations, and a SaaS platform dedicated to monitoring charging and discharging safety. Aihuanhuan is committed to establishing a new generation of intelligent battery swapping ecosystems.

The company provides safe, convenient, and highly efficient battery swapping services for electric motorcycles, electric mopeds, and electric bicycles, thereby promoting urban energy development and enhancing travel experiences. Its comprehensive solution encompasses battery research and development alongside energy storage systems, as well as the infrastructure of battery swapping stations supported by a robust SaaS platform for safety monitoring during charging and discharging.

For B-end users (business clients), Aihuanhuan has introduced the "unlimited mileage and electricity" smooth riding package; while for C-end users (individual consumers), multiple basic packages are available. These packages include a fixed amount of electricity with any usage exceeding this limit charged according to actual consumption.

Regarding compatible models, vehicles wishing to join Aihuanhuan's battery swapping network must be custom-produced and successfully pass the company's rigorous safety verification process.

6.2.1.2 Passenger Cars and Commercial Vehicles

A battery swapping operator is defined as an enterprise or institution that specializes in providing battery swapping services for electric vehicles. By establishing battery swapping stations, these operators offer users the ability to quickly replace depleted batteries, thereby facilitating rapid energy replenishment for vehicles.

There are primarily three types of battery swapping operators. First, there are vehicle manufacturer-led operators: these include companies such as NIO and Geely that either build their own battery swapping stations or collaborate with others to do so. Second, third-party operators focus exclusively on providing battery swapping services; notable examples include Aulton New Energy and the State Grid Corporation of China. Thirdly, there are supplier-led operators: this category includes battery manufacturers like CATL that engage in the provision of battery swapping services.

As of the end of 2023, China has seen a significant increase in the number of battery swapping operators, reaching several hundred entities. The total number of nationwide battery swapping stations has surpassed 10,000. Among these enterprises, NIO and Aulton New Energy hold prominent positions within the market. It is conservatively estimated that by 2025, the total number of battery swapping stations in China will exceed 30,000, indicating substantial potential for market development.

1. **NIO**
   NIO is recognized as one of China's foremost manufacturers of smart electric vehicles and is also among the pioneers in the battery swapping service sector. As of June 2024, seven automotive companies—including Changan Automobile, GAC Group, Geely Holding Group, Chery Automobile, Jianghuai Automobile Group, Lotus Cars, and FAW Group—have established strategic partnerships with NIO to advance battery swapping initiatives collaboratively.
   Business model: NIO has introduced the battery as a service (BaaS) model, enabling customers to acquire electric vehicles without the necessity of purchasing batteries outright. Instead, users can access batteries through a subscription service.
   Strategic positioning: NIO is dedicated to developing a comprehensive nationwide battery swapping network that enhances user experience while simultaneously lowering the barriers to entry for electric vehicle ownership.
2. **Aulton New Energy**
   Aulton New Energy is an enterprise dedicated to the research, development, and operation of electric vehicle battery swapping technology.
   Business model: Aulton New Energy collaborates with various vehicle manufacturers to provide standardized battery swapping solutions tailored to their needs.
   Strategic positioning: Aulton New Energy aspires to become one of the largest battery swapping operators globally, spearheading the advancement of the battery swapping industry through technological innovation and service optimization.
3. **State Grid Corporation of China (SGCC)**
   Introduction: The State Grid Corporation of China is one of the largest power supply enterprises in China and has recently ventured into electric vehicle charging and battery swapping services.
   Business model: Leveraging its extensive network resources, SGCC establishes multiple battery swapping stations aimed at providing convenient services for electric vehicle users.
   Strategic positioning: The SGCC seeks to foster the growth of the new energy vehicle industry by integrating power resources with battery swapping services while reinforcing its leadership position within the energy sector.
4. **Energiex New Energy Technology Co., Ltd.**
   Energiex New Energy Technology Co., Ltd. is a joint venture dedicated to the power battery leasing business, jointly established by Sinopec, PetroChina, SAIC

Motor Corporation Limited, CATL, and Shanghai International Automobile City Group. The company promotes the development of a comprehensive vehicle-battery separation ecosystem through initiatives such as research and development of battery swapping technology, battery operation management, and big data services. Energiex aims to create a standardized platform committed to delivering comprehensive, neutral, and technologically advanced smart energy solutions for the industry. This initiative seeks to provide new energy vehicle owners with safe, reliable, convenient, and efficient high-quality travel experiences.

In alignment with its principles of neutrality and standardization, Energiex New Energy Technology Co., Ltd. is actively engaged in constructing its business platform. In terms of network infrastructure, leveraging the extensive network of over 50,000 gas stations operated by "two barrels of oil" across the country allows for the enhancement and establishment of comprehensive energy service stations that facilitate both battery swapping and charging services. Regarding vehicle models, Energiex will take the lead in collaborating with SAIC Motor Corporation Limited to introduce battery-swapping models across multiple brands. These models will encompass all categories including SUVs, sedans, MPVs (multi-purpose vehicles), as well as commercial vehicles; they will also cater to various scenarios such as private cars, cruising taxis for ride-hailing services online or offline, business travel needs, urban-rural passenger transport, logistics, and express delivery.

5. **EVOGO**

EVOGO is a wholly-owned subsidiary of CATL, committed to delivering convenient and reliable mobile power solutions and services. The comprehensive modular battery swapping solution comprises three key components: the "battery swapping block," the fast swapping station, and an accompanying APP.

Leveraging a vehicle-battery separation model, EVOGO has pioneered the overall solution for modular battery swapping and introduced its new battery swapping service brand, EVOGO. This innovation transforms batteries into shared assets while addressing three critical challenges faced by electric vehicle users: range anxiety, convenience in energy replenishment, and cost associated with purchase and usage.

The "Choco-SEB (Swapping Electric Block)" is a specially developed battery designed for mass production specifically for shared battery swapping applications. It boasts three defining characteristics: compact size coupled with high energy capacity, flexible combination options, and minimalist design aesthetics. Utilizing CATL's latest cell-to-pack (CTP) technology, the "Choco-SEB" achieves an impressive weight energy density exceeding 160 Wh/kg and a volume energy density surpassing 325 Wh/L. A single unit can provide an approximate cruising range of 200 km and is compatible with 80% of pure electric platforms currently available or expected to be launched within the next three years. When consumers engage in battery swaps, they have the flexibility to select one or more blocks as needed to accommodate varying mileage requirements. Furthermore, the "Choco-SEB" incorporates wireless BMS technology; it features only high-voltage positive and negative interfaces externally—significantly enhancing the reliability of pluggable components.

The fast battery swapping station is characterized by four key features: a compact footprint, rapid circulation, substantial capacity, and all-weather functionality. Each standard station requires only three parking spaces. The battery swapping time for a single block is approximately one minute. The station can accommodate 48 battery swapping blocks, ensuring that users always have access to fully charged blocks for replacement without experiencing long wait times. Additionally, the swapping station can offer versions tailored to various climatic and environmental conditions across different regions.

Building on an innovative modular battery swapping solution, EVOGO has established a new service brand named EVOGO, which embodies three core characteristics: one battery serving multiple vehicles, rental of batteries as needed, and the capability for both recharging and swapping.

In contrast to traditional battery swapping models, EVOGO facilitates two adaptable scenarios under the concept of "one battery for multiple vehicles." The "Choco-SEB (swapping electric block)" is compatible with most passenger cars and logistics vehicles developed on pure electric platforms. Furthermore, the swapping station can be customized to accommodate various brands that utilize the "Choco-SEB," effectively overcoming compatibility barriers between batteries and vehicle models while allowing users greater freedom in selecting their preferred battery-swapping options.

Consumers are empowered to "rent batteries as needed," enabling them to choose the number of battery swapping blocks they wish to rent based on their individual usage scenarios and preferences. Currently, in typical daily usage situations, most private car owners utilize only 10–20% of their vehicle's total battery power; however, due to concerns about range anxiety and energy replenishment challenges, they often feel compelled to invest in models with larger battery capacities at significant sunk costs. By adopting a model where consumers can "rent batteries as needed," individuals need only rent one block for daily commuting within urban environments. If they want to travel long distances, they can choose two or three battery blocks, truly achieving increase or decrease as needed and realizing freedom in electricity usage.

At the same time, the EVOGO modular battery swapping solution is both rechargeable and interchangeable, collectively forming a comprehensive energy replenishment strategy alongside existing home charging and fast charging options. This innovation offers enhanced flexibility in energy replenishment.

6. **Qiyuan Green Power**

Shanghai Qiyuan Green Power Technology Co., Ltd. is a leading technology developer and energy service provider specializing in the field of "green electric transportation." As a key implementation unit for the pilot initiative aimed at establishing a robust transportation infrastructure, the company focuses on commercial transportation scenarios, including heavy trucks, construction machinery, ships, and agricultural equipment. Through its self-developed vehicle-storage shared battery systems, operation of charging and swapping facilities, and innovative digital solutions, Shanghai Qiyuan actively promotes the integrated development of transportation and energy sectors.

The company masters core battery swapping patent technologies and pioneered the vehicle-battery separation model for heavy trucks. It has comprehensively laid out more than 800 heavy truck battery swapping stations in all provincial administrative regions except Hong Kong, Macao and Taiwan across the country, forming a nearly 30,000 km charging and swapping trunk line network. It is compatible with more than 83% of battery swapping heavy truck models on the market and serves nearly 30,000 heavy trucks and construction machinery in nearly 100 mines, steel mills, ports, power plants, urban muck transportation, commercial concrete, and trunk line transportation customers or scenarios. It is also committed to promoting China's green electric transportation solution to the world and has laid out business in many countries and regions such as Indonesia.

Urban-level battery swapping service network—mobile heavy truck battery swapping station—mobile battery swapping station—battery swapping mode (spicqyxdl.com).

7. **Enneagon Energy**

Enneagon Energy (enneagon.com).

Shanghai Enneagon Energy Technology Co., Ltd. was established and registered in Shanghai in 2014. The company is recognized as a "specialized, refined, unique, and new" enterprise within the city, serving as a technology center for enterprises in Shanghai and designated as a "little giant" enterprise in Jiading District. It specializes in the research, development, and production of innovative charging and battery swapping equipment for new energy electric vehicles, alongside operating charging stations and related services.

The company emphasizes cutting-edge products and system solutions for electric vehicle charging and swapping while developing an ecological cloud platform to support the operational service network of electric vehicles based on IoT technologies. Its product range includes intelligent onboard terminals, AC/DC charging piles, heavy truck battery swapping stations, as well as mobile IoT/cloud platforms.

By fostering interconnectivity among people, vehicles, equipment, and cloud platforms, Enneagon aims to establish itself as a leading manufacturer of equipment and provider of operation services within the new energy vehicle sector. In 2022 alone, Enneagon's heavy truck battery swapping stations surpassed 200 units; its battery assembly boxes exceeded 30,000 sets; additionally, over 400 models of new energy heavy trucks were announced during this period.

8. **Hangzhou Boton Technology**

Boton Technology is a pioneer in constructing the commercial ecosystem and technical framework for "vehicle-battery separation and modular battery swapping." Through its independently developed "standard box power batteries" and innovative "modular battery swapping method," the company has achieved compatibility in battery swapping across various electric vehicle models, as well as high-level cascade utilization and recycling of power batteries. The concept of "vehicle-battery separation and modular battery swapping" currently represents a robust battery swapping ecosystem with significant practical application.

Boton Technology has meticulously designed and defined a standard battery box that is compatible with multiple vehicle types, including passenger cars, logistics vehicles, and buses. This design facilitates seamless battery swapping within the system. By employing technologies such as "single-pack charging" and "balanced charging" at the battery swapping stations, the standard battery box significantly enhances the service life of power batteries while enabling their cascade utilization.

Furthermore, Boton Technology has engineered vehicle battery swapping assemblies tailored to various vehicle models, alongside developing a central controller known as VCS. Utilizing advanced technologies like blockchain ensures effective interconnection between vehicles and batteries following vehicle-battery separation during modular battery swaps.

In addition to these innovations, Boton Technology has introduced newly designed manual, semi-automatic, and fully automatic battery swapping stations along with dedicated storage solutions for swapped batteries. The company has also independently developed the core control system (GCS) for these stations to guarantee optimal functional operation within the entire network of battery swaps.

The technical system and commercial ecosystem of "vehicle-battery separation and modular battery swapping" developed by Boton Technology has successfully collaborated with seven original equipment manufacturers to create over 10 vehicle models. These products encompass both passenger cars and commercial vehicles.

Currently, Boton Technology has facilitated the deployment of more than 30,000 new energy vehicles equipped with "modular battery swapping" capabilities. The company has provided a total of over 8.87 million battery swapping services, resulting in an accumulated driving distance exceeding 1.34 billion kilometers through this innovative service.

The market implementation of the "vehicle-battery separation and modular battery swapping" business model has not only enhanced user experience but also generated significant economic value for its users.

9. **XJ Electric Co., Ltd.**

XJ Electric Co., Ltd. was originally established as the Xuchang Relay Factory. In 1993, the factory underwent a restructuring process and transitioned into XJ Electric Co., Ltd. through exclusive sponsorship and private placement initiatives. Subsequently, in 1996, the Henan Provincial Economic System Reform Commission issued document number Yu Guopi Zi [1996] No. 107 titled "Reply on the Reconfirmation of XJ Electric Co., Ltd.," which acknowledged that XJ Electric complies with the requirements set forth by the Company Law and officially recognized it as a joint stock limited company. On April 18, 1997, the company was formally listed on the Shenzhen Stock Exchange under the stock abbreviation "XJ Electric" and stock code "000400."

As a leading enterprise within China's power equipment sector, XJ Electric is dedicated to delivering high-end technical equipment and comprehensive system solutions for energy and power to support national economic growth and social development. The company provides extensive technical expertise, product offerings, and service support across all stages of clean energy production, transmission, distribution, and efficient utilization.

XJ Electric focuses on five core business areas: ultra-high voltage technology; smart grid solutions; new energy applications; electric vehicle charging and swapping systems; rail transit innovations; as well as industrial intelligence advancements. Additionally, it explores emerging sectors such as integrated energy services; advanced energy storage technologies; intelligent operation and maintenance practices; along with developments in power IoT applications.

The company's products are extensively utilized throughout various segments of the power system infrastructure. They can be categorized into six primary groups: intelligent power distribution systems; direct current transmission systems; intelligent medium-voltage power supply and consumption devices; smart electricity meters; electric vehicle intelligent charging and swapping systems; alongside EMS processing services.

XJ Battery Swapping Technology Co., Ltd. was established with joint funding from XJ Electric Co., Ltd., China Southern Power Grid Industrial Investment Group, China Southern Power Grid Energy Innovation Equity Investment Fund, and Xuchang Zhichenghe. The company was officially registered on January 31, 2023. Its primary focus is on the research and development of automation equipment, automatic control systems, mechanical devices, mechatronic products, as well as charging and battery swapping systems and facilities. Additionally, it engages in the management of battery systems and oversees the investment, construction, and operation of charging stations for power storage and release.

Leveraging the industrial expertise of China Southern Power Grid and XJ Electric Co., Ltd., the company aims to comprehensively advance product research and development within the battery swapping sector. By integrating advantageous resources from various stakeholders to develop, invest in, construct, and operate projects related to battery swapping technology, it seeks to provide users with comprehensive solutions while fostering a robust ecosystem for the battery swapping industry.

Over the next three to five years, XJ Battery Swapping Technology Co., Ltd. is committed to becoming a leader in electric heavy truck battery swapping solutions by continuously enhancing its investment in technological research and development. The goal is to innovate battery swapping station equipment along with grid-interactive products that meet or exceed national technical standards.

XJ Battery Swapping: Committed to leading advancements in electric heavy truck battery swapping solutions—Discover more about XJ Battery Swapping at Shifair.com

6.2.1.3 Electric Ships

1. **CATL**
   CATL is recognized as one of the world's foremost providers of power battery systems and a trailblazer in electric vehicle battery technology. The company is actively engaged in developing power solutions for electric vessels, offering a comprehensive suite of services that includes battery packs, energy management systems, and battery swapping solutions. CATL supplies high-performance battery packs essential for electric ships and can tailor customized battery swapping service plans to meet customer needs.
2. **Wärtsilä Corporation**
   Wärtsilä stands as a global leader in energy and marine solutions, with its headquarters located in Finland. The company is dedicated to sustainable development and offers advanced ship power systems and technologies. Wärtsilä provides an array of solutions encompassing electric and hybrid propulsion systems while also possessing the capability to deliver battery swapping and maintenance services.
3. **Yara International ASA**
   Yara is a Norwegian fertilizer manufacturer renowned for owning the world's first fully electric zero-emission container ship, "Yara Birkeland." In its initiative involving electric shipping, Yara has implemented an automated battery swapping system designed to facilitate rapid and efficient energy replenishment.
4. **Eco Marine Power**
   Eco Marine Power specializes in research and development focused on green maritime technologies alongside renewable energy applications. The company offers a range of environmentally friendly technologies and products, including solar and wind power generation systems as well as battery energy storage solutions. Additionally, they provide related battery swapping services tailored for electric vessels.
5. **ABB Group**
   ABB is recognized globally as a leading technology innovator across various sectors including electrical products, robotics, motion control, industrial automation, and power grids. For the shipping industry specifically, ABB delivers diverse advanced power automation solutions along with service offerings that encompass both power systems and battery swapping services for electric ships.

6.2.2 Electric Two- and Three-wheeler Manufacturers

In the battery swapping industry chain, manufacturers of two- and three-wheeled electric vehicles occupy a central position, responsible for producing and selling vehicles that support the battery swapping model. These manufacturers not only fabricate the vehicles themselves but also collaborate with upstream battery suppliers and battery swapping equipment producers to ensure compatibility with established battery swapping standards. Furthermore, some manufacturers actively participate in the construction and operation of battery swapping stations.

1. **Yadea**
   Yadea Electric Vehicles: As a leading brand within the industry, Yadea boasts a 20-year development history characterized by exceptional quality and reliable after-sales service. The company emphasizes a balanced approach to power performance and configuration, holding approximately 20% market share. Yadea primarily offers E2Ws (including electric bicycles and scooters), along with batteries, chargers for these vehicles, and various other components related to E2Ws. In 2023, operating income from E2Ws constituted 71% of total revenue, reflecting a year-on-year growth rate of 12%. From 2019 to 2023, the compound annual growth rate (CAGR) reached an impressive 28%, indicating significant expansion. With evolving national standards, electric bicycles have gradually supplanted electric scooters as the predominant category driving the company's revenue streams. In response to diverse consumer needs in 2023, Yadea broadened its product portfolio while enhancing marketing initiatives and expanding its sales network. Consequently, both its E2Ws and supporting products experienced steady growth.

   After the promulgation of the 2018 version of the "new national standard," Yadea has consistently expanded its distribution channels. From 2019 to 2022, the number of distributors increased from 2,154 to 4,041, reflecting a compound annual growth rate (CAGR) of 23.31%. Similarly, the number of sales points surged from 12,000 to approximately 32,000, achieving a CAGR of 38.67%. By the end of 2022, Yadea's domestic distribution network encompassed nearly all administrative regions in China and comprised a total of 4,041 distributors along with their sub-distributors. With over 32,000 sales points—significantly surpassing those of its closest competitor Emma (1,900/30,000+)—Yadea stands out as a leader among brands within the same industry.

   Currently operating under an effective strategy characterized by multiple merchants on each street and numerous stores within towns has allowed Yadea to exceed more than 40,000 sales outlets. This continuous expansion of its store system enhances Yadea's competitiveness in offline sales channels and enables it to gradually widen its lead over other brands in becoming a prominent name in the E2W sector.

   In terms of product innovation, Yadea has launched the "electric new species" Aihuanhuan series electric vehicles. These models are compatible with the Aihuanhuan intelligent battery swapping platform and are designed specifically to address users' charging challenges.
2. **Luyuan Group**
   The Luyuan Group brand was established in 1997 and is headquartered in Jinhua, Zhejiang Province. Since its inception, it has maintained a leading position within the industry. Luyuan Group Holdings (Cayman) Limited commenced operations in 2009 and was listed on the Hong Kong Stock Exchange in October 2023. It stands as a prominent enterprise in the Chinese E2W sector.

   As the first national high-tech enterprise within this industry, the company is dedicated to spearheading advancements in core technologies for electric vehicles. Its product offerings encompass three primary categories: electric bicycles, electric mopeds, and electric motorcycles. Notably, the company's liquid-cooled electric vehicle series features an innovative liquid-cooled motor supported by nine essential technologies. This series boasts an impressive cruising range of up to 100 km and comes with a quality guarantee of ten years, thereby promoting the evolution of E2Ws towards durable goods.

Leveraging a comprehensive and efficient sales network, Luyuan's products have successfully penetrated markets across 37 countries including Thailand, Indonesia, and the Philippines. Some examples of their electric motorcycles are presented in Figure 6.17.

Battery swapping models: Launched customized Hellobike battery swapping electric vehicles. These models support rapid battery swapping services, which help improve user convenience.

3. **Hellobike**

   Hellobike is a shared mobility service provider. It has jointly launched the Hellobike battery swapping service with CATL and Ant Financial.

   Battery swapping models: Hellobike provides multiple two-wheel electric vehicles that support the Hellobike battery swapping service, covering different price ranges and meeting the needs of different users.

4. **Evoke**

   Evoke focuses on providing electric motorcycles and intelligent battery swapping services.

   Battery swapping models: Provides multiple electric motorcycles that support intelligent battery swapping, such as models U1 and U2.

5. **Gogoro**

   Gogoro is a Taiwanese electric motorcycle manufacturer known for its intelligent battery exchange system. Gogoro was established in Taiwan, China, on August 29, 2011. The founder is Horace Luke. It is an innovative enterprise in the field of electric vehicle battery swapping. Gogoro has also made layouts in the mainland market. In the mainland, Gogoro Network is called Aihuanhuan. At present, Yadea also supports battery swapping services for some models. Gogoro has established the PBGN (Powered by Gogoro Network) smart electric motorcycle alliance. Some models of its member departments use the Gogoro

MQN7 1,500 W LFP battery high speed electric motorcycle

High powerful 800 W motor LED screen electric scooter MYC

MTB high powerful high speed motorcycle electric motorcycle

Figure 6.17 Comparative display of various models of electric motorcycles. *Source*: Luyuan (www.luyuanev.com).

Network battery intelligent exchange service and use Gogoro's portable shared batteries. Gogoro began to deepen its layout in the Southeast Asian market in 2020. In 2021, Gogoro and Gojek cooperated to start battery swapping stations. Last year, it also announced the joint construction of a charging and battery swapping system in Indonesia with Hon Hai (Foxconn), Indonesia Battery Corporation (IBC), and Indika Energy.

Battery swapping models: Provide multiple electric motorcycles that support the Gogoro Network battery exchange service, such as Gogoro 2 and Gogoro 3.

6.2.3 Passenger EV Manufacturers

The construction of the electric vehicle energy replenishment system encompasses essential infrastructure, including battery swapping stations and their associated service systems. The research and development of core equipment for these battery swapping stations necessitates technological advancements in critical components such as charging systems, thermal management systems, quick-change structures, motors, and electronic controls. Additionally, the energy replenishment equipment involved in research and development and manufacturing comprises over a thousand different parts and components sourced from hundreds of upstream suppliers.

Integration of upstream and downstream industrial chains

The manufacturing process for electric vehicle energy replenishment equipment involves several stages: upstream raw material supply, midstream equipment manufacturing, and downstream service operations. This multi-faceted process encourages relevant enterprises to cluster geographically, thereby facilitating the integration of upstream and downstream industrial chains. In the Yangtze River Delta region—characterized by its robust industrial foundation in electronic information materials, mechanical manufacturing, and automation technology—favorable conditions are created for the seamless integration of the industrial chain related to electric vehicle energy replenishment equipment.

Pull on the upstream raw material supply chain

Battery materials: The demand for enhanced battery performance in energy replenishment equipment is on the rise, which will stimulate the mining and processing of battery materials such as lithium, cobalt, and nickel. This trend also encourages advancements in material science research and development, thereby driving the growth of upstream mineral resources and chemical products. A notable representative in this field is CATL, a leading company in battery research and manufacturing.

Electronic components: Energy replenishment equipment necessitates a substantial quantity of electronic components, including semiconductor devices and power electronic components. This requirement fosters the advancement of design and manufacturing industries related to these electronic components as well as printed circuit board assembly (PCBA).

Enhancement of midstream equipment manufacturing

Technological innovation: The core technologies underpinning energy replenishment equipment encompass efficient power conversion technology, along with connector and cable technology. The ongoing research, development, and manufacturing processes associated with these technologies demand continuous innovation, thus promoting technological progress within midstream manufacturing.

Manufacturing capacity: The mass production of energy replenishment equipment requires sophisticated manufacturing processes coupled with high levels of automation. This necessity will drive the manufacturing industry towards achieving elevated technological standards alongside increased automation capabilities.

Expansion of downstream services and operations

Installation and maintenance: The installation process, followed by ongoing maintenance for energy replenishment equipment, calls for specialized service teams. Consequently, this creates employment opportunities within related service sectors.

Operation management: The management of energy replenishment facilities necessitates a comprehensive suite of software systems and management tools, thereby fostering the growth of the information technology service industry.

Industrial synergy and cluster effect

The manufacturing and research & development (R&D) of energy replenishment equipment require not only the involvement of traditional manufacturing enterprises but also collaboration across various sectors, including software development, network communication, and big data analysis. This cross-industry cooperation facilitates resource sharing and technological integration among diverse industries, resulting in an industrial synergy effect.

Technology research and development and talent gathering

The core technologies associated with electric vehicle energy replenishment equipment encompass high-efficiency batteries, fast charging technology, battery swapping technology, and intelligent charging networks. Advancing these technologies demands a concentration of highly skilled R&D personnel as well as specialized research institutions, leading to the formation of innovation clusters centered on research activities.

The establishment of industrial clusters is frequently accompanied by talent aggregation and knowledge sharing. The advanced technological nature inherent in the manufacturing processes and R&D efforts for electric vehicle energy replenishment equipment attracts a significant influx of engineers, researchers, and managerial talents to cluster regions. Concurrently, close collaborations between enterprises as well as interactions between businesses and scientific research institutions facilitate the rapid dissemination of knowledge and technology while enhancing overall R&D efficiency.

The position of passenger car manufacturers in the battery swapping industry chain

In the battery swapping industry chain, passenger car manufacturers primarily occupy a central position, responsible for producing and selling vehicles that support the battery swapping model. These manufacturers not only focus on vehicle production but also collaborate with upstream battery suppliers and battery swapping equipment manufacturers to ensure compatibility with established battery swapping standards. Furthermore, some manufacturers engage in the construction and operation of battery swapping stations to develop their own comprehensive battery swapping networks.

1. **NIO**
   NIO holds a significant share in the high-end electric vehicle market. Its innovative battery swapping model offers users a convenient method for energy replenishment, attracting numerous consumers. As the network of battery swapping stations continues to expand, NIO's sales are steadily increasing. Currently, NIO has established an extensive array of battery swapping stations across multiple cities nationwide to provide users with efficient and rapid service. For instance, in major urban areas, users can complete a battery swap within just a few minutes, significantly reducing downtime. Simultaneously, NIO is consistently launching new models and services aimed at enhancing user experience and further driving market sales. Models such as ES6, ES8, EC6, ET5, and ET7 all support the innovative battery swapping mode. The outward appearances of ES6 and ES8 are shown in Figures 6.18 and 6.19.
2. **Geely Auto**
   Ruilan Automobile is a brand under Geely Auto that focuses on new energy battery swapping travel and is committed to providing users with efficient and convenient travel solutions. Ruilan Automobile is gradually emerging in the new energy market with its battery swapping models. Its models are mainly aimed at the travel market and some private users, and are attracting attention for their high cost performance and convenient battery swapping services. With Geely's continuous investment and promotion in the field of new energy, Ruilan Automobile's market share is also gradually expanding. At present, Ruilan Automobile is actively expanding sales channels, strengthening cooperation with partners, and enhancing brand awareness and market competitiveness. Battery swapping models: Maple 80V, Maple 60S, Ruilan Maple 80v PRO, Ruilan Maple 60s PRO, Ruilan 9, etc. Figures 6.20 and 6.21 have demonstrated two of them.
3. **SAIC Motor Corporation Limited**
   SAIC Motor Corporation Limited (abbreviated as SAIC Motor) is a prominent automobile manufacturing enterprise in China, renowned for its extensive experience in automotive research and development, production, and sales. Brands such as Marvel R under SAIC Motor are dedicated to the creation of intelligent electric vehicles, aiming to provide users with high-quality travel experiences.

Figure 6.18 ES6. *Source*: NIO (https://app.nio.com).

Figure 6.19 ES8 (https://www.meta-autolife.com).

Figure 6.20 Ruilan X3 PRO (Geely Auto – Zhejiang Geely Holding Group).

Figure 6.21 Maple 80 V (Geely Auto – Zhejiang Geely Holding Group).

SAIC Motor enjoys a strong reputation in the new energy vehicle market, bolstered by its significant brand influence and technological prowess. The introduction of battery swapping models offers users enhanced choices. Currently, SAIC Motor is expediting the construction and deployment of battery swapping stations while collaborating with Sinopec and PetroChina to establish comprehensive energy service stations that facilitate both battery swapping and charging.

With the ongoing enhancement of the battery swapping network, it is anticipated that SAIC Motor's battery swapping models will achieve improved sales performance in the market. The Marvel R7 will be among the first models introduced under this innovative battery swapping system. Additionally, brands

Figure 6.22 SAIC Maxus Mifa 7 (mifa7phev.njianzhan.com).

such as Roewe, MG, and Maxus are set to successively launch their own battery swapping models across various categories, including SUVs, sedans, MPVs, and commercial vehicles.

Battery Swapping Models: - SAIC Feifan R7 - SAIC Maxus Mifa 7 (Figure 6.22) - SAIC Maxus Mifa 9

Passenger car manufacturers play a crucial role in the battery swapping industry chain. They must not only ensure that their vehicles are compatible with the battery swapping model but also collaborate closely with both upstream and downstream enterprises to jointly advance the development of battery swapping technology.

Cooperation with battery suppliers: It is essential to guarantee that the batteries utilized in vehicles meet established battery swapping standards, thereby enhancing compatibility and interchangeability among different battery types.

Collaboration with battery swapping station operators: Manufacturers should work together to develop an extensive network of battery swapping stations, providing users with convenient access to efficient battery replacement services.

Participation in standard setting: Engaging in the formulation of standardized protocols for battery swapping alongside other manufacturers and industry organizations is vital for promoting uniformity within the sector.

By fostering collaboration with upstream and downstream partners, passenger car manufacturers can enhance their product competitiveness while simultaneously driving progress across the entire battery swapping industry chain. Ultimately, this will enable them to offer users more convenient and efficient energy supply solutions.

6.2.4 Commercial Vehicle Manufacturers

1. **Overview of commercial vehicle manufacturers**
   Battery swapping commercial vehicle manufacturers represent a crucial link within the battery swapping industry. These manufacturers concentrate on the production and research of electric commercial vehicles designed for rapid battery exchange, catering to the specific requirements of sectors such as logistics, public transportation, and long-distance transport. Through technological innovation, they deliver efficient and environmentally sustainable transportation solutions while advancing the electrification and intelligent transformation of the commercial vehicle sector.

   As illustrated in Figure 6.17, commercial vehicle manufacturers occupy a central position within the supply chain of battery swapping entities, situated in the midstream segment of the industrial chain. They must not only maintain close collaboration with upstream partners—namely battery suppliers and manufacturers of battery swapping equipment—to ensure both supply continuity and technical compatibility but also engage with operators of battery swapping stations in order to facilitate seamless integration between vehicles and these stations.

   The end users located downstream in this industrial chain are direct beneficiaries of products offered by commercial vehicle manufacturers. By supplying high-quality battery-swapping commercial vehicles, these manufacturers assist end users in reducing operational costs, enhancing transportation efficiency, and minimizing environmental impact. Additionally, energy storage systems play a vital role in effective battery management and cascade utilization; they collaborate with providers specializing in battery swapping applications to optimize comprehensive lifecycle management for batteries.

   Policy and standard formulation provide normative guidance and policy support for commercial vehicle manufacturers, ensuring the safety, reliability, and universality of their products. Technological innovation and business model evolution serve as significant driving forces behind industry development. Manufacturers are continually exploring advancements in technology and innovative business models to enhance product competitiveness and adapt to shifting market dynamics.

   Market forecasts offer manufacturers valuable insights into emerging trends, enabling them to adjust production plans, sales strategies, and product development directions in alignment with anticipated market demands.

   In summary, battery swapping commercial vehicle manufacturers play a pivotal role within the industrial chain, which is illustrated in Figure 6.23. Through close collaboration with various stakeholders, they not only facilitate the electrification transformation of the commercial vehicle sector but also provide robust support toward achieving goals related to green, intelligent, and efficient transportation. As technology advances and markets mature, battery swapping commercial vehicle manufacturers are poised to assume an even more critical position in the future landscape of the commercial vehicle market.

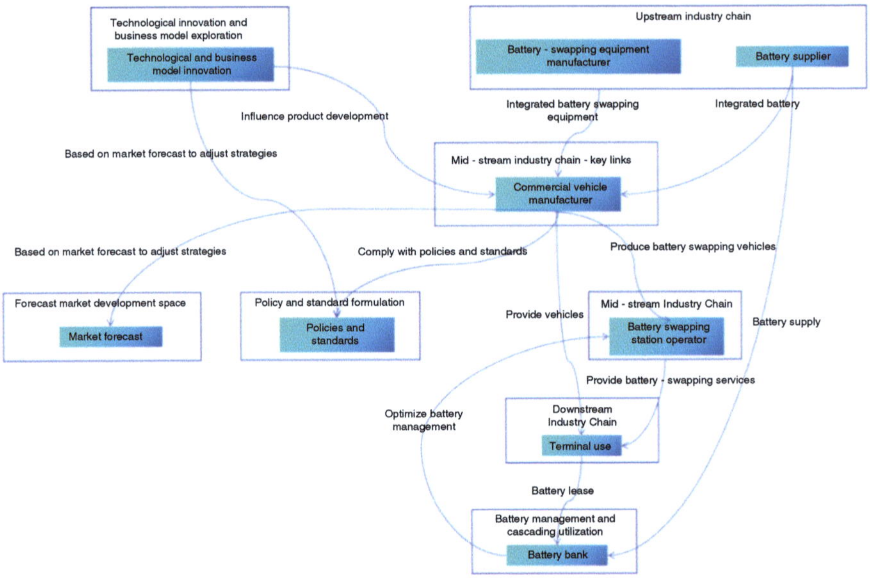

Figure 6.23 Industrial chain map of battery swapping commercial vehicle manufacturers.

2. **Major commercial vehicle manufacturers**

In the light truck field, JAC Motors (JAC) is a well-known commercial vehicle manufacturer in China. Its battery swapping light trucks have an early layout, fast model updates, and a wide range of product categories. At the same time, it cooperates deeply with DeepWay and takes into account the research and development, and sales of battery swapping heavy trucks. The main battery swapping light trucks are as shown in Figure 6.24. The model adopts a Petrel 170-horsepower motor, is equipped with a 90 kWh battery from Ningde City, has a cruising range of more than 220 km, takes 1.5 hours to charge, and takes 3 minutes to swap batteries.

NEWRIZON is a new force brand in the pure electric field established in 2020. As shown in Figure 6.25, in 2023, it launched the iC1 all-in-one charging and swapping pure electric light truck. It adopts chassis battery swapping technology and has a drag coefficient of 0.399. It is mainly sold in regions such as the Middle East. At present, the models launched by NEWRIZON are pure electric light truck models, and will be expanded to the field of new energy heavy trucks in the future.

In the heavy truck field, XCMG Group is at the top of China's battery swapping heavy truck sales from January to June 2024. Battery swapping models include tractors, dump trucks, and mixer trucks, and all are rear-mounted battery swapping. As shown in Figure 6.26, it is a new generation of lightweight battery swapping tractor model developed by XCMG Group. The vehicle's own weight is reduced by more than 500 kg compared to similar products. At the same time, it has a series braking energy recovery system to improve the braking energy recovery efficiency and lower power consumption.

Figure 6.24 JAC all-in-one charging and swapping light truck–Junling Treasures Basin EVs.

Figure 6.25 NEWRIZON iC1 all-in-one charging and swapping pure electric light truck.

DeepWay uses a relatively novel chassis battery swapping technology and adopts an all-in-one charging and swapping energy replenishment mode. Its main model is the battery swapping tractor. As shown in Figure 6.21, the Xingchen model adopts a streamlined front, and the drag coefficient is reduced from 0.65 of the traditional flat head to 0.35. The overall vehicle adopts a chassis battery swapping

Figure 6.26 XCMG Group's new generation of lightweight battery swapping tractor.

mode, and the cargo box volume is increased by 9.6%. Currently, it has assisted driving functions, and will complete L4-level autonomous driving functions in the future. DeepWay began to be put into the Chinese market in the second half of 2023. As of May 2024, the delivery volume has exceeded 1,000 vehicles [1].

6.2.5 Electric Maritime and Aviation

In the battery swapping industry chain, manufacturers of battery swapping ships occupy a central position. Their upstream partners primarily include battery suppliers, manufacturers of battery swapping equipment, and suppliers of ship components, all of whom provide essential batteries, equipment for battery swapping, and various parts for ships. The downstream segment consists of shipping enterprises and companies that operate these vessels, responsible for their operation and utilization.

Below are some key manufacturers in the field of battery swapping ships:

China Innovation Aviation Established on December 8, 2015, this company is headquartered at No. 1 Jiangdong Avenue in Jintan District, Changzhou City, Jiangsu Province. The legal representative is Liu Jingyu, with a registered capital amounting to 1,772,301,858 yuan. It is recognized as a large state-owned enterprise and high-tech entity that houses a national enterprise technology center. As of 2023, the company employs approximately 593 individuals.

The organization primarily focuses on research and development (R&D), production, sales activities as well as market application development related to lithium-ion power batteries; it also specializes in battery management systems, energy storage batteries, integrated products, and lithium battery materials. Its offerings find extensive applications across diverse sectors including new energy passenger vehicles, commercial vehicles, energy storage markets, and specialized application markets. For instance, it supplies power batteries for new energy vehicles along with energy storage system products tailored for power plants and grid enterprises, while also providing customized battery solutions for specialized equipment such as sightseeing vessels.

It possesses continuous capabilities for technological innovation. Leveraging multiple scientific research platforms, it consistently invests efforts in material technology, structural technology, manufacturing technology, and innovations related to ecological and healthy development. Its battery products boast advantages such as high power output, extended cruising range, enhanced safety features, and prolonged service life, thereby demonstrating strong competitiveness within the industry. For instance, the replaceable marine container-type battery products supplied for "COSCO Shipping Lvshui 01" have enabled the vessel to achieve full-electric navigation, showcasing its technological prowess in the domain of electric ship power batteries.

China State Shipbuilding Corporation Limited: As a major state-owned enterprise and a key player in China's shipbuilding sector, China State Shipbuilding Corporation holds a significant position within the global shipbuilding arena. It is equipped with robust R&D capabilities alongside comprehensive design and manufacturing expertise, as well as engineering general contracting abilities. Its numerous subsidiaries span various fields including shipbuilding, ship design, and ancillary services—collectively forming an integrated ship industry chain.

China State Shipbuilding Corporation (Hong Kong) Shipping Leasing Co., Ltd.: This entity boasts extensive experience coupled with substantial financial strength in investment and financing for maritime projects. It provides stable financial support for shipping initiatives to ensure their seamless progression.

Shanghai Merchant Ship Design and Research Institute: Renowned as a leading research institution within China's naval architecture field, it houses a professional design team complemented by advanced design technologies. The institute offers high-quality design solutions tailored to diverse needs and application scenarios while possessing profound technical expertise in overall ship design, structural engineering, and power system configuration.

Wuchang Shipbuilding Heavy Industry Group Co., Ltd. and Wuhu Shipyard Co., Ltd.: As prominent shipbuilding enterprises, they possess advanced shipbuilding facilities and exceptional craftsmanship. These companies are capable of executing shipbuilding tasks with high quality in accordance with design specifications, demonstrating strong competitiveness in both construction quality and timelines.

SAIC Motor Corporation Limited: SAIC Motor is a leading domestic automobile enterprise group with extensive technical expertise and rich production experience in the automotive manufacturing sector. The company boasts a robust research and development team, state-of-the-art production equipment, and a comprehensive supply chain system. It holds a significant market share and exerts considerable brand influence within the automobile industry.

In the realm of marine power batteries, its subsidiaries, SAIC Times Power Battery System Co., Ltd. and Shanghai Econtrol Power System Co., Ltd., leverage SAIC Motor's technological advantages and resource capabilities to actively expand their business operations in the field of marine power batteries. These two entities have invested substantial resources into battery research, development, and manufacturing. They possess formidable technical R&D capabilities as well as production capacities that enable them to deliver high-performance, reliable power battery products for battery-swapping vessel projects.

The company has garnered valuable experience through its involvement in the batch inland river pure electric zero-carbon container ship project based in Shanghai. By collaborating with other enterprises, it continuously enhances product performance and service quality while contributing to the promotion of green development within inland river shipping.

CSSC Saisi Electric Technology (Fujian) Co., Ltd.: As a leading enterprise in China's ship electric propulsion system sector, the company boasts mature technology and extensive experience in both ship electric propulsion systems and marine battery systems. Its products are characterized by high stability and reliability, effectively meeting the power propulsion requirements of various types of vessels.

The company is equipped with professional production facilities and adheres to a stringent quality control system to ensure product quality and production efficiency. Located in Lianjiang, its production base encompasses a plant area of 5,266 square meters alongside an office space of 690 square meters. Plans are underway to establish a battery pack production line, an assembly area for frequency conversion electric control equipment, as well as testing areas for frequency converters and battery factory operations, all complemented by relevant production testing equipment.

Since its inception, CSSC Saisi has successfully signed multiple projects, including the development of a 1,000-ton battery-swapping cargo ship in Huzhou, Zhejiang Province, as well as providing PSV solutions for new energy power propulsion in Singapore. The company has garnered positive results and established a strong reputation within both domestic and international markets. Furthermore, it is continuously expanding its business scope while strengthening collaborations with other enterprises to enhance market competitiveness.

Shanghai Qiyuan Green Power Technology Co., Ltd.: This company actively engages in collaborative projects related to battery swapping ships in Shanghai. It has conducted comprehensive research into business models that separate electricity from maritime operations along with box-type battery swapping technologies. Through ongoing innovation and optimization of technical solutions, Shanghai Qiyuan has significantly improved the operational efficiency and economic viability of battery swapping ships while contributing to advancements in this emerging technology sector.

It collaborates with various entities, including Shanghai Hantang Shipping Co., Ltd. and China State Shipbuilding Corporation (Hong Kong) Shipping Leasing Co., Ltd., to establish a robust cooperation network. By integrating resources from all stakeholders, it actively promotes the green development of inland river shipping and has garnered significant reputation and influence within the industry.

As the demand for sustainable practices in inland river shipping continues to rise, Shanghai Qiyuan Green Power Technology Co., Ltd. is well-positioned for growth in the battery-swapping vessel sector. The company will persist in enhancing technological innovation and collaboration, delivering higher-quality products and services to its customers while contributing to the sustainable advancement of inland river shipping.

Yinson GreenTech (YGT): This is the green technology division of Yinson Holdings Berhad, based in Malaysia, established in 2020. Yinson Holdings operates as a global energy infrastructure and technology firm across 18 countries. YGT aims to create a clean, comprehensive ecosystem enhanced by technology that spans shipping, transportation, and infrastructure through investments in new technologies, research and development initiatives, and strategic partnerships.

Current strategic investments include advanced hydrofoil systems for electric vessels, electric bicycles with replaceable batteries, autonomous driving technologies for both electric vehicles and ships, marine energy storage solutions, as well as electric vehicle charging infrastructures. YGT is dedicated to developing an exclusive digital marketplace that offers affordable, low-carbon technology products and services designed to assist enterprises and communities in achieving their net-zero emission objectives.

In the development of Singapore's inaugural fully electric cargo ship, Hydromover2, YGT plays a pivotal role as a key cooperation partner. The company actively engages in promoting and implementing the project while providing support for the application of green technologies and the research and development of battery systems for the vessel.

Seatech Solutions (Seatech), established in 2000, is a globally operating professional ship design and engineering firm. With extensive experience and expertise in ship design, engineering, and consulting, Seatech has earned an excellent reputation within the industry. Headquartered in Singapore, it maintains branches across various countries and regions, including China, India, Myanmar, Thailand, Japan, Vietnam, Malaysia, South Korea, Australia, and Germany.

As a center of excellence in maritime and marine engineering fields, Seatech boasts a multidisciplinary team capable of leveraging cutting-edge modern technologies to deliver high-quality yet cost-effective services to its clients. Its business scope encompasses all facets of ship design—including engineering analysis and technical consulting—and offers customized ship design solutions tailored to client needs.

In Singapore's first fully electric cargo ship initiative, Seatech serves as the project leader responsible for overseeing overall design and engineering implementation. Furthermore, Seatech leads the Goal Zero Consortium dedicated to advancing zero-emission transformations at Singapore's ports while providing technical support and innovative solutions for electric vessel development.

Goal Zero Consortium (GoalZero): Led by Seatech Solutions, this consortium was established in response to the maritime decarbonization initiative advocated by the Maritime and Port Authority of Singapore (MPA) and the Singapore Maritime Institute (SMI). Its primary objective is to facilitate Singapore's transformation into a zero-emission port. By engaging in research, testing, and deployment of electric ship solutions, it aims to achieve sustainable development within the port industry.

The consortium comprises numerous enterprises and research institutions across various sectors, including shipbuilding, battery technology, and charging infrastructure, thereby forming a comprehensive industrial chain cooperation alliance. In addition to Seatech Solutions and Yinson GreenTech, it includes other companies with specialized expertise and experience in the maritime domain that collectively support project implementation.

GoalZero provides extensive technical support and solutions for Singapore's first fully electric cargo vessel, Hydromover2. This encompasses aspects such as ship design, battery system research and development, as well as planning for charging facilities. Through collaboration among its members, GoalZero addresses challenges related to technology integration, operational efficiency, and commercial viability of electric vessels while establishing a solid foundation for their promotion and application.

Wärtsilä Corporation: Headquartered in Finland, Wärtsilä is a leading global provider of energy solutions and marine technologies with a rich history spanning 190 years. Employing approximately 17,800 individuals across 79 countries worldwide, its business operations are diverse, offering innovative technologies along with lifecycle solutions tailored for both the shipping industry and energy markets.

In the energy sector, we offer flexible and cutting-edge solutions and services designed to manage and optimize complex energy systems. This includes advanced fuel-driven balanced engine power plants, hybrid power solutions, as well as energy storage and optimization technologies. In the marine industry, we provide a comprehensive range of ship technologies encompassing engines, propulsion systems, hybrid power solutions, data analytics tools, liquid and gas processing capabilities onboard vessels, and integrated power systems.

Advantages in the realm of electric ships: We are committed to sustainable development and possess advanced technology along with extensive experience in electric and hybrid propulsion systems. Our offerings include high-performance and reliable power systems coupled with technical support for electric vessels. Additionally, we have the capability to provide battery swapping services and maintenance support that ensure seamless operation of electric ships.

Eco Marine Power is a company dedicated to researching and developing green ship technologies alongside renewable energy applications. It demonstrates exceptional professionalism and innovation within the field of environmental protection technologies and products.

It provides a series of environmental protection technologies and products, including solar and wind power generation systems. These innovative solutions can deliver supplementary power support for electric ships while enhancing their energy utilization efficiency as well as cruising range. Furthermore, we also provide battery energy storage solutions tailored to meet the needs of electric vessels, ensuring stable operational performance at sea. In addition to this, our company offers battery swapping services related specifically to electric ships, which facilitate their efficient operation.

It continuously explores and develops innovative technologies and solutions aimed at enhancing the performance and environmental sustainability of electric ships. For instance, by optimizing the battery management system, both the service life and charging efficiency of batteries are significantly improved. Additionally, new energy conversion technologies are being developed to enhance the utilization efficiency of solar and wind energy.

6.2.6 Drone and Aircraft Manufacturers

Joby Aviation: In January 2022, Joby Aviation's second preproduction prototype received special airworthiness certification from the Federal Aviation Administration (FAA) of the United States and airworthiness approval from the United States Air Force.

In March 2022, it completed system and compliance reviews. The compliance review evaluated software and airborne electronic hardware development and verification, while the system review assessed aerospace grade system and equipment development methods, both approved by the FAA.

In May 2022, it obtained a Part 135 Air Carrier Certificate from the FAA, enabling on-demand commercial air-taxi services.

In August 2022, it expanded its contract with the US Department of Defense. Various military branches, including the Marine Corps, Army, Navy, and Air Force, will test their eVTOL for applications like emergency medical response.

In October 2022, it applied for certification with the Japan Civil Aviation Bureau. Delta Air Lines became a long-term partner, investing $60 million with a potential increase to $200 million.

Lilium: In May 2022, Lilium successfully completed its second Design Organization Approval (DOA) audit, focusing on type certification of its Lilium Jet with the European Aviation Safety Agency (EASA) and the FAA.

In July 2022, four companies (the Bristow Group, AAP Aviation, ASL Group, and Helicopter Airlines) planned to purchase the Lilium Jet.

In September 2022, its technology demonstrator, Phoenix 2, achieved a full transition from hover to wing-borne flight.

In October 2022, Saudi Arabian Airlines signed a memorandum of understanding to establish an eVTOL network in Saudi Arabia, including purchasing 100 eVTOL aircraft. The company also completed a financing round of $119 million.

Archer Aviation: In April 2022, Archer Aviation founded a joint advisory committee with United Airlines to guide eVTOL maintenance and operation concepts.

In July 2022, it selected Honeywell to provide flight-control actuation and micro VCS thermal-management technologies for its eVTOL aircraft.

In August 2022, it received a $10 million predelivery payment from United Airlines for 100 production-model aircraft and revealed design details of its eVTOL.

In November 2022, it presented the production-model aircraft Midnight to the public. It also announced an eVTOL route with United Airlines connecting Newark Airport and a Manhattan heliport. Its demonstrator achieved the transition from vertical lift to full wing-borne flight.

Volocopter: In February 2022, Aviation Capital Group (ACG) agreed to collaborate with Volocopter on financing solutions for post certification sales, covering up to $1 billion worth of Volocopter eVTOL aircraft.

In May 2022, its four-seat fixed wing eVTOL, Volo Connect, completed its first flight, with potential service entry as early as 2026.

Airbus urban air mobility: Airbus is developing the CityAirbus eVTOL concept. It has a quadcopter configuration with four large-sized rotors for vertical lift, suitable for short-distance urban passenger transport. Airbus has conducted strict tests, including successful unmanned test flights, demonstrating the concept's feasibility.

Geely: In 2017, Geely acquired Terrafugia, entering the flying car and eVTOL sectors. It has since made several strategic moves, including investing in Volocopter, founding a joint-venture company, and launching the self-developed AE200 project.

Autoflight: Its R&D team, based in Shanghai, focuses on manned eVTOL research and development. In 2021, it completed a financing round of $100 million.

6.3 Downstream Market Applications and Services

6.3.1 B2B Users

1. **Electric two- and three-wheelers**
   The battery swapping service for electric two- and three-wheelers is not universally adopted due to its high costs. B-end consumers tend to be more market-oriented, driven by their pursuit of efficiency. In contrast, the consumption habits of C-end consumers—who typically purchase vehicles while renting batteries—have yet to fully develop. Consequently, it has become an inevitable path for the advancement of battery swapping that B-end stakeholders take the lead in cultivating C-end demand.

   To facilitate the penetration of battery swapping among C-end consumers, numerous manufacturers are entering the market through a multi-party cooperation model. This approach not only expands market reach but also mitigates risks associated with unilateral investments by leveraging strong partnerships, thereby achieving stable and healthy business growth. A notable example is the "Aihuanhuan Intelligent Battery Swapping Platform," which has been jointly launched by Yadea Technology Group, Dachangjiang Group, and Gogoro.
2. **Passenger vehicles and commercial vehicles**
   The B-end users in the battery swapping industry primarily refer to commercial entities, specifically organizations or enterprises that incorporate electric vehicles into their business operations. These users typically have heightened demands for operational efficiency and cost management, making the battery swapping model particularly advantageous for them.

   The B-end market is experiencing rapid development, with battery swapping heavy trucks entering a phase of accelerated growth. In contrast to C-end battery swapping solutions, the B-end market—encompassing taxis/ride-hailing services and heavy trucks—has entered a period of swift commercialization. As of November 2022, BAIC Group has deployed a total of 32,000 battery swapping taxis in Beijing and established 175 battery-swapping stations, with 115 currently operational. Presently, various vehicle usage scenarios—including trunk logistics vehicles, muck trucks, tractors, and mining trucks—are actively advancing the adoption of the battery swapping model; related products and their commercialization are gradually being validated by market demand.

In 2022, sales figures for new energy heavy trucks saw significant growth as battery-swapping heavy trucks increasingly emerged as the mainstream option within this segment. During the first eleven months of 2022 alone, cumulative sales of new energy heavy trucks reached 19,312 units—a remarkable year-on-year increase of 157.91%. The penetration rate rose from just 0.56% during the same period last year to an impressive 3.12%. Notably among these figures were cumulative sales of battery-swapping heavy trucks totaling 9,082 units—a staggering year-on-year increase of 265.03%, accounting for approximately 47.03% of all new energy heavy truck sales.

In addition, driven by the increasing demand for energy conservation and emission reduction, the electrification of construction machinery has commenced. Major construction machinery manufacturers have successively introduced electrified products into their portfolios. Companies such as XCMG, Shandong Lingong, and Liugong have explored and developed battery-swapping construction machinery solutions, launching various battery-swapping products in the market. For instance, XCMG's XCH908E2 stacker is primarily utilized for loading and unloading empty containers at ports, site transfers, and stacking operations. This model enables rapid battery swapping within just three minutes. Compared to traditional internal combustion stackers, it offers an annual cost savings exceeding 300,000 yuan.

The sales distribution of different models of battery-swapping heavy trucks aligns with current scenarios where these vehicles are predominantly employed in short-distance transportation and port transfer activities. The advantages of ultra-fast energy replenishment coupled with low operating costs inherent to the battery-swapping model effectively meet operational demands for high efficiency and cost-effectiveness among vehicle operators. Consequently, this has facilitated a phase of volume expansion for battery-swapping heavy trucks.

The following outlines several common categories of B-end users within the battery swapping industry chain:

Taxi and online car-hailing companies: Taxi companies and online ride-hailing service providers represent some of the early adopters of the battery swapping model. For these enterprises, maximizing vehicle uptime and operational efficiency is paramount. The implementation of a battery swapping system significantly reduces charging wait times for drivers—thereby enhancing vehicle utilization rates and overall revenue generation.

Logistics and delivery companies: Logistics companies require efficient goods transportation solutions to maintain operational effectiveness. The adoption of a battery swapping model ensures that electric trucks or lorries can quickly return to full charge status in minimal timeframes—thus avoiding delays associated with prolonged charging periods while ensuring timely deliveries.

Public transport operators: Including bus companies and urban light rail systems, these operators must ensure the continuous operation of their vehicles. Establishing battery swapping stations at key routes or hubs allows buses to quickly exchange batteries en route, thereby maintaining uninterrupted service.

Heavy truck transportation industry: In long-distance transportation and heavy cargo handling, the battery swapping model can significantly reduce charging times. This is particularly crucial for heavy trucks that require prolonged operational periods, ultimately enhancing transportation efficiency and lowering operating costs.

Enterprise fleets: Many enterprises maintain their own fleets for employee commuting, business travel, or goods delivery. The battery swapping model enables these fleets to sustain high vehicle availability while simultaneously reducing maintenance expenses and minimizing environmental impacts.

Municipal and public service departments: Public service vehicles—such as those used in postal services, waste collection, and road maintenance—can also benefit from the battery swapping model. This is especially advantageous in urban settings where frequent short trips are necessary.

Construction and mining: In these sectors, implementing a battery swapping system for electric construction machinery and vehicles ensures continuity of operations at construction sites or mines. It helps avoid work interruptions caused by waiting for charging.

The value of the battery swapping model for B-end users lies in its ability to assist businesses in adapting to the trend of electrification transformation while meeting increasingly stringent environmental protection standards. By reducing vehicle downtime, improving operational efficiency, lowering overall operating costs, and providing more flexible energy management solutions, this model proves beneficial. As technology matures and infrastructure improves further, the application of the battery swapping model within the B-end market will become increasingly widespread.

6.3.2 Consumer (C) End-users

The C-end users within the battery swapping industry chain, specifically referring to end consumers (consumer-to-consumer, C2C), encompass individual consumers or household users who utilize electric vehicles in their daily lives and opt for battery swapping services as a means of replenishing their vehicle's energy.

As the electric vehicle market continues to expand, the demand for battery swapping services among C-end users is steadily increasing, particularly in the sectors of two-wheeled and four-wheeled electric vehicles.

The promotion and profitability threshold for C-end battery swapping is relatively high, and the market remains in an early investment phase aimed at capturing market share. There exists a wide variety of private car brand models, each with differing battery pack specifications and configurations. The scenarios for energy replenishment demands are diverse, as are the geographical areas involved. To effectively promote the battery swapping model, it is essential to establish a comprehensive network of battery swapping stations across cities. However, challenges such as inconsistent standards, substantial investment costs, and difficulties in achieving profitability have somewhat hindered the development of C-end battery swapping.

Among automobile manufacturers, only NIO has taken a proactive approach by focusing on the C-end battery swapping market. Since launching the ES8 in 2017, all electric vehicles produced by NIO have been equipped with support for battery swapping technology.

Since 2022, both SAIC and Geely have sequentially entered the consumer battery swapping market. In the latter half of 2022, Geely's Ruilan Automobile launched a series of battery swapping models, including the Ruilan 7, Ruilan Maple 80v PRO, Ruilan Maple 60s PRO, and Ruilan 9. In September 2022, SAIC Motor Corporation Limited collaborated with CATL, Sinopec, PetroChina, and Shanghai International Automobile City to jointly invest in and establish Shanghai Energiex New Energy Technology Co., Ltd.

In terms of network infrastructure, Energiex New Energy Technology Co., Ltd. leverages Sinopec's and PetroChina's extensive network of over 50,000 gas stations nationwide to upgrade existing facilities into comprehensive energy service stations capable of supporting both battery swapping and charging services.

Regarding vehicle models, brands such as Marvel R, Roewe, MG, and Maxus are set to introduce various battery swapping models across all categories, including SUVs, sedans, MPVs (multi-purpose vehicles), and commercial vehicles. The Marvel R7 will be among the first models offered under this new battery swapping initiative. Additionally, GAC Group is also planning to launch its own battery swapping models in 2023.

The following outlines some typical characteristics and application scenarios of C-end users within the battery swapping industry chain:

Two-wheel electric vehicle users: This group represents the earliest and most prevalent category of C-end users for battery swapping services, encompassing electric motorcycles and bicycles. Given that the battery capacity of two-wheeled electric vehicles is relatively limited, battery swapping services can swiftly restore their operational capability, making them particularly well-suited for short-distance travel in urban environments.

Owners of private electric vehicles may opt for battery swapping services, especially during long-distance journeys or in emergency situations. In instances where charging stations are unavailable or when charging times are excessively lengthy, battery swapping offers a rapid means of energy replenishment, thereby minimizing waiting periods.

Electric car sharing users: Participants in time-sharing rental or car-sharing programs may also emerge as C-end users of battery swapping services. This service model prioritizes immediate availability and efficient turnover of vehicles; thus, the implementation of a battery swapping system ensures that cars remain consistently fully charged.

For C-end users, the primary value proposition of battery swapping services lies in their convenience and time efficiency. In contrast to traditional charging methods, which can be time-consuming, battery swaps can be completed within minutes—significantly reducing user wait times while lessening reliance on existing charging infrastructure. However, factors such as the cost associated with battery swapping services, coverage rates for swap stations, and standardization issues regarding batteries also play crucial roles in influencing decisions among C-end users. As technology advances and market conditions mature, these challenges are progressively being addressed; consequently, the appeal of battery swapping services to C-end consumers continues to grow steadily.

6.3.3 Aftermarket Services and Retired Battery Recycling Utilization

Post-battery swapping market services and the recycling and utilization of retired batteries are critical components in the new energy vehicle industry chain. These aspects not only impact the daily experiences and operational costs for electric vehicle users but also have a direct influence on long-term environmental protection goals and resource recycling efforts.

The turnover and management of batteries within the battery swapping model offer distinct advantages to enterprises, particularly regarding the efficient recycling and circular utilization of power batteries.

This battery swapping model allows automakers to maintain direct oversight over the entire lifecycle of batteries—from production, usage, maintenance, to recycling—thereby creating a closed-loop supply chain. Such centralized management enhances both the efficiency of battery recycling processes and enables companies to more accurately monitor battery performance degradation, facilitating timely maintenance or replacement. This approach ensures optimal battery health as well as stable cruising capabilities for electric vehicles.

As the network for battery swapping expands, enterprises can manage retired batteries on a larger scale. Through centralized collection, testing, and classification processes, batch processing of these batteries becomes feasible, leading to reduced costs associated with processing individual units. Furthermore, large-scale battery recycling fosters long-term collaborative relationships between enterprises and recycling manufacturers, further driving down overall costs.

Battery swapping automakers that effectively manage the complete lifecycle data of batteries are better positioned to conduct performance analyses and foster technological innovations in battery cascade utilization. Although retired batteries may no longer be suitable for vehicle use, they still possess sufficient capacity to fulfill the requirements of application scenarios such as energy storage systems and backup power supplies. By repurposing these batteries for fixed energy storage projects, automakers can create new revenue streams while minimizing resource waste and mitigating environmental impacts.

For consumers, having assurance that the batteries utilized in their electric vehicles are professionally maintained and have a clear recycling pathway enhances their confidence in purchasing electric vehicles. The battery health monitoring and prompt replacement services offered under the battery swapping model alleviate users' concerns regarding battery degradation and residual value, thereby increasing the overall appeal of electric vehicles.

As global regulations on battery recycling and environmental protection become increasingly stringent, enterprises engaged in battery swapping with efficient recycling mechanisms can more readily comply with policy requirements while avoiding potential legal risks. Furthermore, this model underscores an enterprise's commitment to sustainable development, enhancing its brand image and social responsibility.

In the long term, efficient recycling and reuse capabilities for power batteries will emerge as a significant competitive advantage for automakers utilizing battery swapping technology. This approach not only diminishes reliance on raw materials and reduces production costs but also creates new market opportunities for enterprises, such as battery leasing services and energy storage solution providers. Consequently, this positions them more favorably within the burgeoning new energy vehicle industry.

The battery swapping model not only demonstrates clear advantages in terms of battery recycling and circular utilization but also offers multifaceted competitive benefits to automakers. These include cost optimization, technological innovation, enhanced consumer trust, and improved corporate social responsibility.

Several major market participants are introduced below:

Brunp: A leading battery recycling enterprise in China, cooperating with CATL.

Guangdong Brunp Recycling Technology Co., Ltd. (hereinafter referred to as Brunp Recycling) has swiftly emerged as a leading enterprise in the global waste battery recycling sector since its inception in 2005. The company specializes in the recycling and resource utilization of waste batteries, extracting essential metal elements such as nickel, cobalt, and manganese from these batteries for reapplication in the production of battery materials, particularly ternary cathode precursor materials. This process not only facilitates efficient resource recycling but also significantly reduces reliance on primary mineral resources while promoting sustainable environmental development.

In 2015, CATL, a global powerhouse in power battery manufacturing, strategically invested in Brunp Recycling and established it as a holding subsidiary. This strategic move further enhanced CATL's control over the battery material supply chain and enabled comprehensive closed-loop management spanning from battery manufacturing to waste battery recycling. The synergy between Brunp Recycling's technical expertise and CATL's industrial resources fosters deep integration across upstream and downstream industrial chains, thereby bolstering overall industry competitiveness.

Brunp Recycling has established seven production bases worldwide that encompass the entire process—from waste battery collection and disassembly to extraction and new material production. Concurrently, the company has developed five scientific research and innovation platforms dedicated to continuous investment in research and development. It actively deepens strategic collaborations with both domestic and international enterprises along the industrial chain as well as renowned scientific research institutions—constantly driving technological innovation and facilitating industrial advancement.

As of now, Brunp Recycling has achieved an annual waste battery processing capacity of 120,000 tons. Utilizing its innovative directional recycling technology, the company has effectively addressed the challenge of "waste reduction" in the waste battery recycling process and attained a remarkable recovery rate of nickel, cobalt, and manganese metals at 99.3%. This accomplishment underscores Brunp Recycling's leadership position in the global waste battery recycling sector and significantly contributes to advancing the global circular economy and green energy revolution.

The success story of Brunp Recycling not only highlights the technical prowess and innovative capabilities of Chinese enterprises in the realm of waste battery recycling and utilization but also establishes a benchmark for the international waste battery recycling industry. It exemplifies China's commitment and proactive measures in tackling global climate change while striving for sustainable low-carbon development.

Umicore: A Belgian materials technology and recycling company, excelling in precious metal and battery material recycling.

Umicore, a leading global materials technology group, operates across three core business areas: catalysis, energy and surface treatment technologies, and recycling. The company is at the forefront of clean technology, with a significant portion of its revenue and research and development investments concentrated in this domain—particularly in catalysts for emission control, rechargeable battery materials, and innovative recycling technologies. Umicore's strategic plan outlines an investment of nearly 5 billion euros from 2022 to 2026 in phases, emphasizing its focus on the battery recycling sector and rechargeable battery materials. This commitment underscores the company's dedication to fostering a sustainable energy future.

Umicore holds a pioneering position in the field of battery recycling. Its facility, located in Hoboken, Belgium, has been operational since 2011 and possesses the capacity to process 7,000 tons of waste batteries annually equivalent to approximately 35,000 electric vehicle battery packs. In February 2022, Umicore introduced its next-generation proprietary recycling technology. This optimized recycling process ensures that recovered metal materials achieve battery-grade purity and can be directly utilized in the production of new lithium batteries, thereby realizing a true closed-loop material cycle.

To further enhance its position in the battery recycling market, Umicore has entered into a cooperation agreement with Automotive Cells Company (ACC) to provide battery recycling services for its pilot plant located in Nersac, France. This partnership aims to meet ACC's demand for recycled materials. Additionally, Umicore is planning a significant expansion in the United States and intends to invest 500 million euros to establish a new facility. Utilizing the company's most advanced hydrometallurgical technology, this new plant is expected to recycle 150,000 tons of waste batteries annually. Once operational in 2026, Umicore will become the first company in Europe to achieve large-scale integration within the cathode active material (CAM) value chain, thereby creating a comprehensive circular ecological system that spans from raw material recycling to new material manufacturing.

Umicore's strategic initiatives and ongoing investments not only reinforce its leadership position in both battery recycling and rechargeable battery materials but also reflect the global shift towards green energy and underscore the significance of adopting a circular economy model. Through these efforts, Umicore provides sustainable solutions addressing the challenges posed by battery waste while making substantial contributions toward global energy transition and environmental protection initiatives.

Botree: Lithium battery recycling equipment + solution provider.

Botree Recycling Technologies is committed to providing a comprehensive solution for the recycling of critical materials used in batteries. By developing advanced separation systems and intelligent equipment, we empower battery manufacturers, operators, automotive producers, energy storage companies, and recyclers worldwide to establish a closed-loop system for battery materials. This initiative promotes low-carbon practices and sustainable development of essential resources in new energy.

Botree comprises experienced experts from leading universities, organizations, and companies around the globe. Drawing on extensive research in battery material preparation, critical metal separation and purification, extractant synthesis, and carbon footprint assessment, Botree has successfully implemented industrial-scale applications for lithium-ion battery recycling. Our processes encompass mechanical size reduction, hydrometallurgical recycling, and material regeneration. Furthermore, we have developed recycling methodologies for hydrogen fuel cells and solid-state batteries as well as an innovative short-separation process for laterite nickel ore. We have also established a comprehensive life cycle assessment (LCA) system for electric vehicle batteries that adheres to international standards.

In 2019, Botree became the first Chinese company to participate in the "International Energy Agency Hybrid and Electric Vehicle Technology Collaboration Programme (IEA HEV-TCP) Task 40 Critical Raw Materials for Electric Vehicles (CRM4EV)." In December 2021, it was announced that IEA HEV-TCP Task 48 Battery Swapping was initiated by Botree Recycling Technologies. Additionally, Botree serves as one of the primary authoring units of several key publications including "Blue Book on the Recycling of Spent Power Lithium-Ion Batteries," "Recycling of Power Lithium-Ion Batteries," "R&D Trends in Next Generation Batteries from the Perspective of Leading Chinese Researchers," as well as being convenor of ISO/TC 333/WG9.

Botree offers comprehensive industrial and technical consulting services, along with intelligent mobile recycling equipment capable of processing 500–5,000 tons of battery packs per year. We also design and install large-scale engineering recycling facilities that can handle between 10,000 and 100,000 tons of battery packs annually. Additionally, we assist our clients in establishing a complete system for responsible sourcing, life cycle assessment, carbon emission reduction, and IREC renewable energy certificate trading. Botree has built a robust customer base that includes nearly 10 companies from the Fortune Global 500 list within the new energy sector. By focusing on the recycling of critical battery materials and supporting the circular design of next-generation batteries, Botree Cycling collaborates with stakeholders across the entire industry chain to create a truly sustainable and carbon-neutral future.

Botree Recycling Technologies is committed to providing an all-encompassing solution for new energy battery recycling. Through the development of advanced separation systems, innovative recycling processes, and smart equipment solutions, we empower global battery manufacturers, operators, automotive producers, energy storage enterprises, and recyclers to achieve a closed-loop system for battery materials. This initiative promotes low-carbon practices and supports sustainable development in key resources related to new energy. Our goal is to facilitate your recycling process swiftly while ensuring it operates efficiently and profitably over time.

7

Challenges and Opportunities

7.1 Infrastructure, Cost, and Safety

With the rapid development of the electric vehicle industry, battery swapping has emerged as a novel energy supply method that is gradually garnering market attention. However, despite its advantages—such as swift energy replenishment and prolonged battery life—the battery swapping model still encounters several challenges related to insufficient infrastructure development, high costs, and safety concerns. The main factors that would affect the future industrial prospects of battery swap stations have been summarized in Figure 7.1.

7.1.1 Infrastructure Construction Challenges

Large upfront investment: Establishing a battery swapping station necessitates substantial capital investment, which includes expenses for land leasing or acquisition, equipment procurement and installation, as well as site construction. For instance, GCL Energy Technology's announcement indicates that the construction cost for a passenger vehicle battery swapping station is approximately 2.6 million yuan, while the cost for a heavy truck battery swapping station can reach up to 4.2 million yuan. When factoring in investments for batteries and electrical lines, the total required investment may escalate to around 5 million yuan for passenger vehicles and 10 million yuan for heavy trucks. Companies like NIO have made considerable early-stage investments in building these stations but are consequently facing heightened financial pressures.

Difficult site selection: The selection of an appropriate location for a station involves numerous considerations such as traffic accessibility, support from surrounding power facilities, and proximity to users. In urban areas where suitable land resources are limited and highly contested with elevated costs, conversely, in remote regions where land may be more affordable but local traffic flow and user demand must also be taken into account; otherwise, there exists a risk of underutilization.

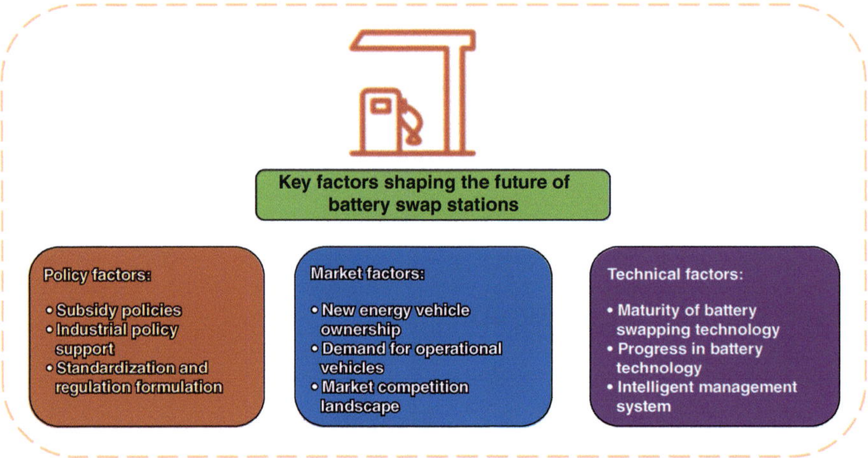

Figure 7.1 The main factors affecting the future industrial prospects of battery swap stations.

Nonuniform standards: Electric vehicles from different manufacturers exhibit variations in battery specifications and interfaces, leading to limited compatibility among battery swapping stations. This lack of standardization complicates the provision of universal services across multiple vehicle models, thereby increasing both construction and operational complexity and costs. For instance, Tesla's chassis battery swapping technology suffers from low compatibility due to the unique shapes of its batteries, which cannot be shared across various vehicle series and models. This limitation restricts operational efficiency and hinders potential development.

Unreasonable layout: Currently, the distribution density and spatial planning of battery swapping stations do not adequately meet user demands. In certain urban core areas, there may be an insufficient number of battery swapping stations, forcing users to travel greater distances when they require a swap—thereby negatively impacting their experience. Conversely, in remote or low-traffic regions, some battery swapping stations may remain underutilized, resulting in resource wastage.

7.1.2 Cost Challenges

High construction cost: Beyond the initial investment for construction, ongoing expenses also encompass equipment maintenance and renewal costs. The machinery at battery swapping stations necessitates regular upkeep, including maintenance checks, overhauls, and upgrades—to ensure safe operation; all these factors contribute to elevated long-term operating costs.

High battery cost: Batteries represent the core asset of the battery swapping model, and their costs constitute a significant portion of overall expenses. Furthermore, as battery technology continues to evolve, both performance and cost are subject to constant change. This dynamic environment presents challenges for battery swapping operators in terms of procurement and inventory management. On one hand, it is essential to ensure that the quality and performance of batteries meet user requirements; on the other hand, there is a need to manage battery inventory judiciously to prevent excess stock and depreciation resulting from rapid advancements in battery technology.

High operating cost: The daily operations of battery swapping stations necessitate substantial human, material, and financial resources. This includes staff salaries, site rental fees, and equipment energy consumption costs, among others. For instance, a flagship battery swapping station for a particular brand aims to serve 1,000 vehicles per day but is projected to take approximately three to four years before reaching the breakeven point. During this period, the pressure from operating costs remains considerable.

Lack of economies of scale: Currently, the adoption rate of the battery swapping model remains low; consequently, the number of users is relatively small, while utilization rates at existing stations are not high enough. This situation hampers efforts to achieve economies of scale that could help reduce unit costs. Only when the number of users reaches a certain threshold can costs be distributed more broadly across operations, thereby enhancing operational efficiency and profitability.

7.1.3 Safety Challenges

Battery safety: The battery swapping process involves frequent disassembly and installation of batteries, which may impact the structural integrity and performance of the batteries. This increases the risk of damage, leakage, overheating, and even fire or explosion. For instance, if the battery interface becomes loose or worn after multiple swaps, it can lead to poor electrical contact, resulting in heating issues and other safety concerns. Furthermore, there are variations in quality and performance among different batteries. Ensuring the safety and reliability of swapped batteries is a critical issue.

Data security: The battery swapping process entails handling substantial amounts of user data and vehicle information, including vehicle identification numbers, battery status details, and user driving patterns. Protecting this data is essential; any breach or malicious use could expose users to risks such as privacy violations or vehicle theft.

Equipment safety: Battery swapping stations utilize complex and highly precise equipment that includes battery-swapping robots, charging systems, monitoring devices, etc. Malfunctions or improper operation of this equipment can result in safety incidents that pose risks to both individuals and vehicles. For example, if a battery-swapping robot mispositions itself during the grabbing or installation process due to mechanical failure or inaccuracies in alignment, it may cause the battery to fall or inflict damage on the vehicle.

Lack of standards and insufficient supervision: Currently, the safety standards and specifications within the battery swapping industry are not yet fully developed, resulting in a lack of a unified safety inspection and certification system. This situation may lead to potential safety vulnerabilities during both the construction and operational phases for certain battery swapping enterprises. Additionally, it poses significant challenges for regulatory authorities, making it difficult to effectively supervise and manage the overall safety of the battery swapping industry.

7.2 Standardization of Battery Swapping

Standardization has always been a key issue hindering the development of battery swapping technology. Its critical aspects include the technical solutions for battery swapping, battery packs, and other components that are crucial to the development of the electric vehicle (EV) battery swapping industry. Standardization enhances system interoperability, ensuring compatibility and interchangeability between batteries and swapping equipment from different manufacturers. This not only provides users with greater freedom of choice but also improves the accessibility of the battery swapping mode.

Moreover, standardization reduces the costs of technology deployment and maintenance, bringing significant benefits to the construction and operation of battery swapping stations. In addition, by establishing safety regulations and testing standards, standardization plays a key role in ensuring the safety and reliability of battery swapping systems. The lack of standardization may lead to fragmentation within the battery swapping industry, ultimately reducing its overall efficiency. This fragmentation may result in different regions or companies adopting different technologies and standards, causing compatibility issues and restricting the mobility and operating range of electric vehicles.

Furthermore, standardization lowers market entry barriers, promotes competition, creates more opportunities for new enterprises to enter the market, and stimulates innovation and development in the industry.

Despite significant progress in the revision of China's battery swapping standards, the essence of battery swapping standardization requires compromise between automotive manufacturers and battery swapping operators. Different stakeholders often have different technical preferences, which makes the process of reaching consensus and coordination complex. Automotive manufacturers and battery swapping operators may prioritize their own interests, making it challenging to reach an agreement on technology and standards. In addition, different local standards and group standards follow different technical paths, resulting in hardware interface sizes and communication protocols that cannot be effectively shared or interchanged.

At present, the battery swapping market faces challenges due to nonuniform standards, leading to compatibility issues between battery swapping equipment and batteries from different brands. However, as industry standards gradually become unified, the convenience of the battery swapping mode is expected to significantly improve, leading to a rapid increase in market penetration. For example, some local governments and industry associations have begun to promote the formulation and implementation of standardization practices for battery swapping of electric two-wheeled vehicles.

7.3 Technological Advancements and Policy Support

The progress of battery swapping technology and policy support has brought many development opportunities to the battery swapping industry, which are mainly reflected in the following aspects:

7.3.1 Development Opportunities Brought by Technological Progress

7.3.1.1 Enhancing Battery Swapping Efficiency

1. **Rapid battery swapping technology**
   Advanced battery swapping equipment and technology enable rapid battery exchanges, significantly reducing the energy replenishment time for vehicles. For instance, some modern battery swapping stations can complete a battery swap in just a few minutes, comparable to the refueling time of traditional fuel-powered vehicles. This capability is essential for enhancing user convenience and satisfaction while attracting more consumers to opt for new energy vehicles.

   The implementation of rapid battery swapping technology broadens the application of new energy vehicles in operational contexts. Vehicles with high efficiency demands—such as taxis, ride-hailing services, and logistics transport—can maintain uninterrupted operations through the battery swapping model, thereby improving vehicle utilization rates and economic benefits.
2. **Automated battery swapping system**
   The automated battery swapping system minimizes manual intervention by streamlining operational processes and enhancing the accuracy and reliability of battery exchanges. The use of robots or automated equipment for removing and installing batteries not only reduces labor costs but also mitigates safety risks associated with human error during operation.

   This automated system can provide continuous service 24 hours a day to accommodate users' diverse battery swapping needs at any given time. Additionally, it monitors both the status of batteries and the performance of the swapping equipment in real-time, allowing for timely fault diagnosis and maintenance that enhances the stability and reliability of battery swapping services.

7.3.1.2 Optimizing Battery Performance

1. **Battery technology innovation**
   With the continuous advancement of battery technology, key performance indicators such as energy density, cruising range, and lifespan have seen significant improvements. High-energy-density batteries offer extended cruising ranges

for vehicles, alleviating users' concerns about range anxiety. Additionally, long-life batteries contribute to reduced operational costs for users and enhance the recycling rates of batteries.

The research, development, and application of innovative battery technologies—such as solid-state batteries and sodium-ion batteries—have created new opportunities for growth within the battery swapping industry. These emerging battery types are characterized by higher safety standards, faster charging capabilities, and lower costs; they are anticipated to further accelerate the evolution of the battery swapping model.

2. **Upgrade of battery management system (BMS)**
An advanced BMS is capable of real-time monitoring of various parameters related to battery status, including voltage, current, temperature, and remaining power—to ensure that operations remain within safe limits. Furthermore, the BMS can facilitate balanced management across multiple cells in a battery pack to extend overall service life.

The deep integration between the BMS and the battery swapping system enables intelligent management solutions for batteries. For instance, by analyzing a vehicle's driving route and power requirements in advance, strategically located battery swapping stations can be planned for users—thereby enhancing both efficiency and accuracy in providing battery swapping services.

7.3.1.3 Expanding Application Scenarios

1. **Adaptation to different vehicle models**
The advancement of battery swapping technology has made it feasible for the battery swapping model to be applied across a wider range of vehicle types, including passenger vehicles, commercial vehicles, and specialized vehicles. For instance, large vehicles such as heavy trucks and buses can achieve rapid energy replenishment through the battery swapping model, thereby enhancing operational efficiency. Additionally, specialized vehicles like sanitation trucks and fire engines can utilize the battery swapping model to ensure uninterrupted operation during emergencies.

The versatility of different vehicle models significantly expands the market potential of the battery swapping industry and creates additional business opportunities for enterprises engaged in this sector. Furthermore, it contributes to the promotion and adoption of new energy vehicles across various fields.

2. **Combination with intelligent network connection technology**
The integration of battery swapping technology with intelligent network connectivity enables seamless intercommunication between vehicles and battery swapping stations. Through intelligent terminals, vehicles can access real-time information regarding the locations, battery inventory status, and queuing details of nearby battery swapping stations. This functionality allows users to make advanced reservations for battery swapping services, thereby minimizing waiting times.

Moreover, intelligent network connectivity facilitates remote monitoring and management of batteries. It enables the timely detection of potential issues with batteries and provides early warnings to enhance safety and reliability.

Additionally, by leveraging big data analytics, the layout and operational management of battery swapping stations can be optimized to improve service quality and efficiency.

7.3.2 Development Opportunities Brought by Policy Support

7.3.2.1 Clarifying the Development Direction

1. **National policy orientation**
   The country has implemented a series of policies aimed at fostering the development of new energy vehicles, with a clear emphasis on promoting innovation and the application of the battery swapping model. For instance, the "New Energy Vehicle Industry Development Plan (2021–2035)" advocates for the acceleration of charging and battery swapping infrastructure construction while encouraging the adoption of the battery swapping model. This provides explicit policy guidance and support for advancing the battery swapping industry.
   Local governments have also responded promptly to national directives by issuing specific implementation rules and subsidy policies designed to promote the development of the battery swapping model within their respective regions. For example, certain cities offer assistance in terms of land allocation and funding for constructing battery swapping stations, as well as providing purchase subsidies and operational incentives for battery swapping vehicles to encourage both enterprises and users to actively engage in this model.
2. **Formulation of industry standards**
   Policy support plays a crucial role in expediting the formulation and enhancement of industry standards related to battery swapping. Establishing unified industry standards can standardize technical requirements, safety performance metrics, and interface specifications for battery swapping equipment. This will enhance compatibility and universality among such equipment while promoting healthy growth within the sector.
   Moreover, developing these industry standards can strengthen oversight over battery-swapping enterprises, thereby improving service quality and safety levels. Simultaneously, it aids in reducing research-and-development costs as well as operational risks faced by companies, ultimately enhancing overall competitiveness within this burgeoning industry.

7.3.2.2 Increasing Capital Investment

1. **Fiscal subsidies**
   The government actively encourages enterprises to enhance their investment in research and development of battery swapping technology, as well as in the construction of related infrastructure, through fiscal subsidies. For instance, a certain percentage of construction and operational subsidies may be allocated to companies that establish battery swapping stations, thereby alleviating the financial burden associated with initial investments and ongoing operations.
   Additionally, purchase subsidies will be provided to consumers who acquire battery swapping vehicles, aimed at increasing user acceptance and intention to

purchase such vehicles. These fiscal incentives can effectively stimulate market demand and facilitate the rapid advancement of the battery swapping industry.

2. **Financing support**

   Policy initiatives are designed to create more financing channels and opportunities for enterprises within the battery swapping sector. The government can guide financial institutions to enhance credit support for this industry by offering financial services such as low-interest loans and financing guarantees tailored for these businesses.

   Furthermore, social capital is encouraged to establish investment funds specifically targeting the battery swapping industry, which would support technological innovation and market expansion among relevant enterprises. Such financing support can provide robust financial backing for battery swapping companies while promoting their growth and development in an increasingly competitive landscape.

7.3.2.3 Optimizing the Market Environment

1. **Simplifying the approval process**

   The government is committed to streamlining the approval process for the construction of battery swapping stations, thereby enhancing overall efficiency. For instance, a one-stop approval service platform will be established to consolidate the approval procedures across multiple departments, including land use, urban planning, and environmental protection, with the aim of reducing approval timelines. By simplifying this process, we can minimize both time and operational costs for enterprises, expedite the development of battery swapping stations, and effectively address market demand for battery swapping services.

2. **Strengthening market supervision**

   Policy support plays a crucial role in strengthening market supervision within the battery swapping industry while standardizing market practices. This includes rigorous oversight of quality standards, safety performance, and service benchmarks associated with battery swapping equipment. Additionally, measures will be taken to combat counterfeit products and unfair competitive practices. Robust market supervision safeguards users' legitimate rights and interests while fostering trust and satisfaction in the battery swapping sector. Furthermore, it contributes to establishing a fair, competitive environment that promotes sustainable growth among battery swapping enterprises.

7.4 Future Market Demand

7.4.1 New Energy Vehicles Market

7.4.1.1 Taxi and Ride-hailing Sector

Taxis and ride-hailing vehicles operate for extended periods and cover long distances, requiring high energy replenishment efficiency. With the increasing adoption of new energy taxis and ride-hailing vehicles in various regions, the battery swapping model has emerged as a viable solution to meet their rapid energy replenishment needs.

Demand for battery swapping in this sector is expected to continue growing. For example, some large cities have already begun large-scale replacement of traditional taxis with new energy vehicles and have simultaneously constructed battery swapping stations. In the future, it is anticipated that more cities will take similar measures.

7.4.1.2 Logistics and Transportation Field

Logistics delivery vehicles typically have fixed routes and require frequent energy replenishment to maintain transportation efficiency. Adopting the battery swapping model can significantly reduce vehicle charging waiting times, thereby increasing vehicle utilization rates. This is crucial for logistics companies looking to reduce operational costs. Some well-known logistics companies have already begun experimenting with new energy logistics vehicles using the battery swapping model.

7.4.1.3 Private Passenger Vehicle Field

Although the adoption of the battery swapping model in private passenger vehicles currently faces some challenges, such as nonuniform standards and insufficient battery swapping stations, continuous technological progress, improved standards, and increased consumer awareness of battery swapping may positively influence consumer choices. Many consumers may opt for the battery swapping model due to its convenience and the assurance it provides for battery life. This trend is particularly evident in city centers or along highways, where the demand for rapid energy replenishment is strong. Therefore, the demand for battery swapping in private passenger vehicles is expected to gradually increase.

7.4.2 Electric Two-wheeler Battery Swapping Market

Food delivery and express delivery industry: The food delivery and express delivery sectors impose stringent requirements on the cruising range and energy replenishment speed of electric two-wheelers. The battery swapping model allows riders to replace batteries in a short time, enabling them to resume work promptly, thereby significantly enhancing operational efficiency. As the volume of business in food delivery and express delivery continues to grow, the demand for electric two-wheeler battery swapping services is expected to experience explosive growth. For instance, several battery swapping service providers have introduced tailored solutions specifically designed for the food delivery and express delivery industries, which have received widespread acclaim.

Ordinary resident users: For ordinary residents, electric two-wheelers serve as an essential mode of transportation for daily commuting. While charging remains the predominant method at present, the convenience and flexibility offered by battery swapping also appeal to certain users. This is particularly true in older residential areas where charging infrastructure may be inadequate; here, the battery swapping model effectively addresses users' charging challenges. Additionally, some individuals opt for battery swapping services due to concerns regarding battery longevity. It is anticipated that demand for electric two-wheeler battery swapping among ordinary residents will gradually increase in the future.

7.4.3 Electric Ship Battery Swapping Market

Short term (1–3 years): In the short term, the market demand for electric ship battery swapping is expected to grow gradually. This growth will be driven by the advancement of environmental protection policies and the impetus provided by demonstration projects in certain regions. In various inland river basins and offshore ports, the utilization of electric ships is anticipated to increase steadily. This includes applications such as city ferries, sightseeing vessels, and some small cargo ships. However, the growth rate of market demand in the short term is likely to be relatively moderate. This is primarily due to the current scale of electric ship ownership and the progress made in developing battery swapping infrastructure.

Medium term (3–5 years): In the medium term, the technology associated with electric ships will continue to mature. This will lead to further enhancements in battery performance and a gradual reduction in costs. As a result, more shipping enterprises are likely to focus on the operational efficiency of electric vessels, recognizing the advantages offered by battery swapping models. Notably, within inland river transportation and short-distance coastal transport sectors, there will be a significant rise in market demand for electric ships. For example, several large logistics companies may initiate pilot operations using electric cargo vessels to lower transportation costs while adhering to environmental regulations. Concurrently, local governments may also enhance their investment efforts toward constructing battery swapping infrastructure for electric ships to further stimulate market demand.

Long term (more than 5 years): In the long term, the electric ship battery swapping market is anticipated to experience significant growth. As global demand for environmental protection continues to rise, the pressure on energy conservation and emission reduction within the shipping industry will intensify. Electric ships are poised to become one of the mainstream options for future maritime transport. Large container vessels and ocean-going cargo ships may gradually transition to electric or hybrid power systems, thereby substantially expanding the market demand for electric ship battery swapping. Furthermore, with ongoing technological advancements, key performance indicators such as cruising range and charging speed of electric ships are expected to improve continuously. This will further enhance their market competitiveness and drive rapid growth in market demand.

7.5 Industrial Development Prospects

On December 18 2024, at the Chocolate Swap Battery Ecosystem Conference, Mr. Zeng Yuqun, chairman of CATL (Contemporary Amperex Technology Co., Limited), announced the launch of the new generation of the Chocolate Swap Battery solution and initiated the "Chocolate Swap Battery Ecosystem." This announcement marks a significant step in the standardization of swap battery technology and the construction of infrastructure in the swap battery industry.

The new solution includes two standardized Chocolate Swap Battery blocks (20# and 25#), which are designed for A0 and A/B class electric vehicles, respectively, and are available in both LFP (lithium iron phosphate) and NCM (nickel cobalt manganese) versions. CATL has also collaborated with several automakers to introduce 10 new swap battery models and plans to build a large-scale swap battery network, aiming to construct 1,000 swap stations by 2025, 10,000 stations in the medium term through cooperation with various parties, and ultimately build 30,000 stations through nationwide collaboration. This initiative by CATL signifies important progress in the standardization of swap battery technology and the construction of infrastructure. An example of CATL chocolate battery swap station can be found in Figure 7.2.

Beyond CATL, enterprises across vehicle segments are accelerating swapping deployment, forming a multi-dimensional landscape:

- Commercial vehicles: Up to now, Qiyuan Green Power has deployed more than 1,200 charging and swapping stations for electric heavy-duty trucks across 31 provincial-level administrative regions in China, along with over 100 trunk line networks.
- Passenger vehicles: NIO's alliance has 3,500+ stations (covering 95% of prefecture-level cities) and provides over 86,000,000 swaps in total.

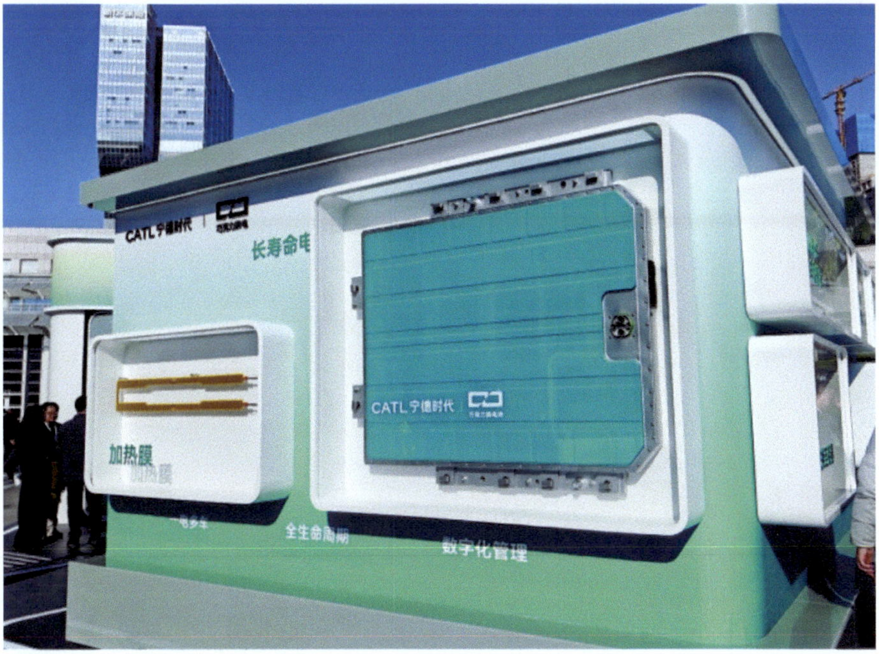

Figure 7.2 Sample of CATL's Chocolate Swap Battery.

Competition between these players drives tech/service upgrades, while cooperation promotes resource sharing and standard unification, benefiting industry health.

Battery swapping is strategically important for optimizing user experience, advancing EV transformation, and achieving "dual carbon" goals. Despite challenges (standardization, costs, profits), it is entering a critical transition from "pilot" to "scale" driven by tech, policy, and demand.

Over the next 5–10 years, China's swapping industry will follow three trends:

1. Standardization as a breakthrough: Unified specs (battery size, interfaces, safety) will enable cross-brand compatibility, lowering barriers.
2. Scenario deepening: Swapping will extend to heavy trucks, engineering machinery, and ships, forming a comprehensive network.
3. Intelligence low carbon integration: AI and big data will enable smart scheduling, while "swap+storage" turns stations into distributed energy hubs, supporting low-carbon energy systems.

In conclusion, while challenges remain, swapping's efficiency advantages and industry collaboration will drive it to become a key force in China's EV industry development.

Index

Note: Page numbers in *italics* and **bold** refers to figures and tables respectively.

c